Artist: Francine Auger

George Mercer Dawson, 1849-1901.

William Chalmers

William Chalmers holds a B.A. degree from the University of Victoria and a Masters degree from Concordia University. He majored in English and Creative Writing. In 1996, he published a novel, *No More Worthy*.

Before he attended university, Chalmers worked in construction, landscaping, aircraft freighting, and fruit picking. He was also a childcare worker, a switchboard operator, and an underground miner. From 1982 to 1989 Chalmers was a Corrections Officer at the Nanaimo Correctional Centre, where he supervised inmates and was a certified fitness tester for the staff.

At present William Chalmers lives in Nanaimo, BC, and works as a head guide for Tofino Expeditions. He leads sea kayak tours on the Central Coast of British Columbia, Haida Gwaii, and Baja, Mexico. He has also worked as an instructor for Brentwood College School in their Outdoor Education Program and for the Malaspina University-College Recreation and Ecotourism Programs.

With his strong interest in wilderness, ecotourism, and the outdoors in general, it is not surprising that William Chalmers chose to write about George Mercer Dawson, the extraordinary "little doctor" from Montreal who surveyed, mapped, and photographed the uncharted wilderness of western Canada in the 1870s and 80s.

In the same collection

George Mercer Dawson

Canadian Cataloguing in Publication Data

Chalmers, William, 1956-

George Mercer Dawson: geologist, scientist, explorer

(The Quest Library ; 8).

ISBN 0-9683601-8-1

1. Dawson, George M., 1849-1901. 2. Geologists – Canada – Biography.
3. Scientist – Canada – Biography. 4. Explorers – Canada – Biography. I. Title.
II. Series.

QE22.D28C42 2000   551'.092   C00-940656-5

Legal Deposit: Third quarter 2000
National Library of Canada
Bibliothèque nationale du Québec

XYZ Publishing acknowledges the support of The Quest Library project by the Canadian Studies Program and the Book Publishing Industry Development Program (BPIDP) of the Department of Canadian Heritage. The opinions expressed do not necessarily reflect the views of the Government of Canada.

The publishers further acknowledge the financial support our publishing program receives from The Canada Council for the Arts, the ministère de la Culture et des Communications du Québec, and the Société de développement des entreprises culturelles.

Chronology and Index: Lynne Bowen
Layout: Édiscript enr.
Cover design: Zirval Design
Cover illustration: Francine Auger

Printed and bound in Canada

XYZ Publishing   Distributed by: General Distribution Services
1781 Saint Hubert Street   325 Humber College Boulevard
Montreal, Quebec   H2L 3Z1   Toronto, Ontario   M9W 7C3
Tel: (514) 525-2170   Tel: (416) 213-1919
Fax: (514) 525-7537   Fax: (416) 213-1917
E-mail: xyzed@mlink.net   E-mail: cservice@genpub.com

DAWSON

George Mercer

GEOLOGIST, SCIENTIST, EXPLORER

XYZ
Publishing

This book is for Mom and Dad.

Contents

Preface		xi
1	1873: Latitude 49	1
2	1874: Where the Buffalo Roam	17
3	A Boyhood Adventurer	33
4	1875: Bushwhacking and Brulé	47
5	1876: Monsters, Tricksters, and Ice	63
6	1877: Salmon and Camels	77
7	1878: To The Charlottes	93
8	1887: Rivers of the Yukon	107
9	1891: A Voyage to the Bering Sea	125
	Epilogue	141
	Chronology of	
	George Mercer Dawson (1849-1901)	147
	Sources Consulted	167
	Index	171

Preface

The primary sources for this project were Dawson's diaries, which he kept through most of his life. Such was his reserve, in all of these texts, that only very occasionally does he display his feelings about people or places in any other than a professional manner. In this regard he is probably a typical man of his time. The challenge in writing this book, then, was to discover the personality of George Dawson, his sense of humour, his fears and his hopes, without speculating or making fiction.

The mandate of the Geological Survey encouraged George to be an active collector of native objects and artifacts, and his field work gave him frequent opportunity to add to the collections being assembled at McGill University and, later, at the Geological Survey of Canada museum in Ottawa. Most of the items he collected were purchased from First Nations people, although on one known occasion he simply gathered up skulls from a burial site without permission. This method was considered acceptable at the time, but today we call it theft. Despite this, George Dawson's contributions to Canadian ethnological collections remain important.

Readers with a background in geology may find fault with George's ideas on glaciation. It is important to read them in the context of the geological knowledge of the late 19th century. For a more detailed discussion of the progression of his theories on glaciation, see Cole and Lockner's "Introduction," in *The Journals of George M. Dawson: British Columbia, 1875-1878,* Vol. 1. See also, H.W. Tipper's GSC Bulletin, *Glacial Morphology.*

In general, Dawson's geological work was of a very high standard. "The acid test of any geologist," writes Morris Zaslow, "is how well his work will stand up when it is re-examined later on the ground by a geologist equipped with new tools and informed by new scientific knowledge. Dawson meets this test better than any man of his generation."

This project was completed with the assistance of several people. Johanne Pelletier of the McGill University Archives, Susan Yates of the Nanaimo Harbourfront Library, the staff at Malaspina University-College Library, John Wilson, Sophie St-Hilaire, and Rhonda Bailey. My thanks to each.

1

1873: Latitude 49

When I think of anybody else getting the
appointment to go survey that splendid
country with splendid scenery, it puts me in
the blues.

George Mercer Dawson looked up from the granite
pebbles by his boot and cocked his head. A sound
like distant surf overrode the low moan of the shoulder-
ing wind. He stood and looked to the westward sky and
saw dark clouds swirling across the prairie towards him.
They seemed to touch the ground and swell up into the
sky as if fanned to life by the wind. George caught up
the reins and spoke to his horse. "Easy boy, easy now."

"He was a li'l hunchback runt – up first in the morning and off over the hills like a he-goat. Wasn't none of us could keep up with him." Officers of the British North American Boundary Survey Commission, 1873. George stands third from right in the back row.

The clouds surged like ocean swells, and as they grew closer, the hissing became a menacing, chattering sound like the gnashing of thousands of tiny teeth.

He knew what this was, now, this breathing, digesting cloud. Quickly, he removed his coat and wrapped it over the horse's head. Holding the horse on a short rein, George pulled his hat down to cover his neck and turned his back to the onslaught. He talked to the horse in a quiet voice, explaining to the animal that the locusts, *Acrididae*, would pass quickly enough, but that there might be little feed for a few days.

The insects flew out of control, pelting George and his horse all over with their bodies and racheting wings. He held the coat firmly over the horse's head and spoke, and when the horse began to step nervously, George moved with the animal, talking to it constantly. For twenty minutes, half an hour, George and his horse stood within the whirling cloud of flying insects. And then it had passed, the hiss and chatter fading into the distance.

George knelt to examine the gnawed remains of fescue grass and vetch, and found egg deposits, which would lie dormant until they hatched out next spring to repeat the cycle. He scraped up several with the blade of his knife and dropped them into an envelope. They were excellent samples.

Locusts: the eighth plague. "For they covered the face of the whole earth, so that the land was darkened; ...and there remained not any green thing in the trees, or in the herbs of the field, through all the land." George could almost hear his father's voice reading aloud the passage.

The feed taken by this swarm would cause hardship not only for George's horse, but also for the several hundred head of horses and oxen belonging to the Boundary Commission Survey.

∞

Ten months earlier, in September 1872, a British and Canadian party of 270 men, 114 horses, 210 oxen, 55 ponies, and 179 wagons assembled at the small border town of Pembina, North Dakota, on the banks of the Red River. Among the members of the party were astronomers, surveyors, topographers, engineers, surgeons, veterinarians, photographers, cooks, tailors, carpenters, wheelwrights, bakers, and blacksmiths. They were joined by an American party, including a detachment of U.S. Cavalry, which nearly doubled their number.

The purpose of the Joint Boundary Commission was to complete the survey and marking of the boundary shared by the United States and Canada – the 49th Parallel – and put to rest any question about where, exactly, the boundary lay.

Beginning late in the season as they did, the surveyors and engineers busied themselves trying to find the "Angle," the northwesternmost point of Lake of the Woods, a boundary point established by David Thompson in 1824. The Northwest Angle was an area that was easy enough to find but which was, upon exploration, found to be little more than a narrow arm of land that petered out into a boggy marsh. With the help of local Métis guides, the surveyors finally located the holes left by the rotted wooden posts from the previous survey, establishing the point from which they would work south to intersect with the 49th parallel.

The American contingent returned home in November, but the Canadians and British members stayed on at Fort Dufferin, Manitoba, through the winter, continuing their work at a reduced pace and making plans for the summer of '73.

That same autumn, young George Dawson entered his final year of studies at the London Royal School of Mines, a year in which he distinguished himself by adding a second Edward Forbes Medal and Book Prize to the Duke of Cornwall's Scholarship and Director's Medal and Book Prize that he had won the previous year. Upon graduating with honours George earned the title of Associate of the Royal School of Mines. And though he could have stayed in England and worked or taught in his field, he had only one thought: get home to Canada and explore the west.

The position of geologist and naturalist on the British-American Boundary Survey wasn't the job George had hoped for. Surveying the prairies held little promise of excitement when he compared it to the land he really wanted to survey: the Rockies and beyond, in British Columbia and the Yukon Territory.

The director of the Geological Survey of Canada, Dr. Selwyn, was prepared to offer George that very position just as soon as he obtained approval from the government minister. But no one knew exactly how long it might take to receive that approval. And, as George's father pointed out, the Boundary Commission job was a very important undertaking for the country, of which George should be proud to be a part.

George knew, of course, that his father was right. But if he should accept the Boundary Commission job and miss the opportunity to be selected for the Geological Survey position in western Canada, he would never forgive himself, or his father. His father was adamant. "Consider boundary decided," read his telegram to George.

Reluctantly, George agreed to take the position. "When I think of anybody else getting the appointment to go to survey that splendid country with splendid scenery," he wrote to his sister Anna, "It puts me in the blues."

∞

So it was that in early June, 1873, George travelled by steamship across Lake Superior to Duluth, Minnesota, by rail to Moorhead, North Dakota, then down the Red River into Canada by river boat and wagon to meet the Canadian Boundary Survey Commission at Fort Dufferin, Manitoba. In his trunks were his photographic equipment – camera, glass exposure plates, processing chemicals, and a small light-proof tent – field notebooks, telescope, compass, barometer, specimen jars and boxes, plant presses, boards and paper and ink, several mousetraps, two wooden chests for bird and animal skins, and his geologist's hammer and canvas sample bags.

The plains weren't exactly the splendid scenery of British Columbia, but it was George's first season in the field, which was a long time in coming after three years of school in England and the best part of his teenage years spent bedridden, bound in body trusses and tormented by severe headaches and pains in his back and limbs. Some days, now, those years seemed like no more than a horrible dream, nearly forgotten, until he overheard someone remark on his height or the shape of his back. But George had no time to dwell on that. There was too much to do.

∞

George awoke with the dawn sliding a wedge of grey light through the crack in his tent door. He emerged from his tent and found himself in the midst of a large camp of men and horses and oxen, all shaking themselves awake under the immense morning sky. Meadowlarks and yellow-headed blackbirds sang morning songs with a sense of urgency, and George hastened to the chuck wagon. The cook, a barrel-chested man of six feet, was bent over two large pots from which he alternately dipped steaming mugs of bitter black tea and large grey dollops of oatmeal. George received his with a grunted thanks, then took his breakfast squatting knee to knee between members of Britain's Royal Engineers and local Métis men.

As he rode away from the wagon train with its lumbering oxen and creaking wagons, George let his eyes sweep across the ocean of grassland before him and realized that this was a land ripe for exploration and inventory, a fine place to sample and classify and put to work his knowledge and training. He trotted his horse to the top of a rise so he could sight up and down the Red River Valley one last time. His empty sample bags slapped lightly against his legs, promising a summer – the first of many, he hoped – of exploration and adventure.

Ten years ago, even five, nobody in his family would've believed that he would ever be able to work in the field, as a geologist or in any other capacity. Nobody except George himself. The illness that had left him a hunchback less than one and a half metres tall had certainly slowed him down, but George knew there was no way he would ever let it keep him from becoming a field geologist and exploring the west.

And here he was, twenty-three years old, a trained geologist, at the edge of the Great Plains. The Rocky Mountains and British Columbia still lay nearly 1300 kilometres away, but George Mercer Dawson's journey towards his goal, his dream, was becoming reality.

Before him the Red River Valley stretched north and south. The valley, with the stories of hundreds of thousands, even millions of years embedded in its banks, was like an uncle with a book in his arms, waiting to be asked for a story. The river flowed north now, though at one time it had flowed south. *There* was a story George would like to sink his teeth into, but for now his stories lay to the west, across the Great Plains. He reined his horse, clucked to it, and began to ride west along the 49th parallel.

∞

As the expedition lumbered westward across the prairies like a giant caterpillar, the Canadian and American parties worked by leap-frogging, and stopping to take exact astronomical observations every thirty-two kilometres. These observations could require up to a week to determine with sufficient accuracy the latitude and longitude. At every 4.8-kilometre interval the men would construct a mound, piling the sod as though building an earthen igloo, to demarcate the Canada-United States Boundary.

This slow pace gave George plenty of time to study the prairie. Rocks and pebbles were abundant, so George began to collect them, his sample bags so full he would often walk back to camp to save weight on his horse. He would then separate the rocks into groups according to mineral composition: granite, limestone, felspar, et cetera, and count the numbers of each type. As he knew where the various types originated, he was then able to deduce the direction of the flow of the ice sheet.

∞

At Turtle Mountain, in southern Manitoba, the monotony of the prairie landscape changed to a broken, hilly region of poplar and white birch groves. Choke cherries, wild roses, and raspberries filled in along the streambeds. Water, grass, and firewood were abundant, but so, too, were mosquitoes, which maddened the livestock and forced the men to wear veils and gauntlets for protection. Only during the midday heat or when a moderate breeze blew was there a break from the incessant hum of the insects.

As George lay in his tent, listening to the mosquitoes beat against the walls like summer rain, he reflected on the afternoon's excitement. An ox, which had been standing in a particularly thick cloud of mosquitoes, had been seized by a coughing fit, then had begun to wheeze laboriously before it stopped breathing altogether. The handlers pounded furiously on the ox's ribs and stomach, trying to induce it to breathe. Finally, the animal collapsed, apparently dead. But the sudden weight on its ribs forced the air out of its lungs one last time and expelled a huge glob of mucous and mosquitoes. The ox wheezed and heaved and started breathing again, to the cheers of George and the handlers. Smudge fires were ordered lit throughout the camp: better to choke on smoke than on mosquitoes. George wrote a short poem that night.

> The air is full of murmur and of song
> That rounds the solemn stillness of the waste
> As gay the light mosquito oars along
> "In God and in his sword" his trust is placed –
> Oh smudge Oh glorious smudge! let me entrench
> in thy sweet noxious cloud
> And nose and eyes all smarting with thy stench,
> there curse the winged crowd!

As the prairie west of Turtle Mountain levelled out, George felt much as he had while crossing the Atlantic ocean aboard a ship: like a man at sea. In every direction ran endless waves of grassland, combed by the travelling wind and interrupted only by the occasional swell of hillock or dip of a coulee. George checked the western horizon frequently, searching for that first blue peak shimmering in the distance.

Often, the only point of reference available was the sun, burning brightly over the open land. At noon George's shadow made a circle at his feet the size of his hat. At sundown he cast a shadow the length of a hundred metres.

They were on the short-grass prairie now, the prairie of the Indian and buffalo. Soon George came upon circles of stones where native people had placed their teepees, and deeply rutted paths that ran in northwest to southeast lines, evidence of thousands of years of buffalo travel.

Large solitary boulders called "erratics" had been dropped by the ice sheet and lay scattered on the plains as if some gigantic hand had used them for a game of marbles. Buffalo paths converged upon the erratics like the spokes of a wheel, and the sides of the boulders were polished smooth by the shoulders and haunches of a hundred thousand rubbing buffalo. The ground at the base was often so rutted by the circling hooves that the boulders seemed raised upon pedestals. George inspected each one carefully, as much amazed by the work of the buffalo as he was by the work of the ice.

The prairies were so much like the sea: at one time a sea of grinding ice; then a great sea of water left by the glacier's retreat; and now, a sea of grass, flawed, scorched and gnawed by wind, fire, and locust.

But it was the ice that George's mind kept coming back to. Ice had transported the boulders and deposited them, without a doubt, but how, exactly? Had the Laurentide glacier pushed its way more than 1100 kilometres westward while gaining some 1067 metres in elevation? It was inconceivable. Or, had the entire plain been submerged beneath the weight of a glacier over 300 metres thick? Or was it the great melting sea that allowed icebergs to carry stones and boulders many miles from their places of origin? These were questions he and his father had debated back in Montreal. And now George rode and walked each day through the landscape of their debates.

∽

In the third week of September, at the equinox, the expedition was struck by a raging snowstorm that lasted for seven days. They corralled the wagons in a horseshoe and lashed canvas tarps to the inside to provide a windbreak. Then the men huddled in their tents inside the makeshift compound. Even the horses, which had been turned loose to graze, chose the protection of the wagons. George worked in his tent, writing and sketching, venturing into the blizzard only at mealtime.

On the eighth day, the members of the expedition crawled from their tents into a world of winter, brilliant beneath a clear blue sky. The snow melted after only a day or two, and three weeks of Indian summer followed the unmistakable warning of winter.

On October 5th, 456 kilometres west of the Red River, the Survey turned back, their season at an end. Making their final survey at Porcupine Creek in southern Saskatchewan, George and the rest of the party packed their gear for the return trip, deciding what to leave cached until next summer and what to take home. George's trunks and bags were full to the brim with plant and soil samples and, of course, geological samples: granite, limestone, felspar, chert, quartz, diorite, hornblende schist, and ironstone.

A week out of Fort Dufferin, George and the Survey crew rode up on a brush fire, which had been caused by a lightning strike. Worried about the wind rising and turning the fire on them, they hurried eastward. About three kilometres on, George heard a shout from the rear: "Wildfire!" He turned in his saddle to see a sky of smoke and ash being rushed toward them by a heat-maddened wind. He raced to the rear of the column and joined the men as they fired a patch of prairie between them and the charging firestorm. They let it burn for a couple of hundred metres, then beat it out and drove all the wagons and livestock onto the burnt patch. The men struggled to control the panicked animals as the air buzzed with heat and a whirling rain of hot ash fell. For half an hour, George and the men fought to calm the animals while the wildfire hissed and crackled across the prairie.

They travelled the rest of that day across a blackened, smouldering plain. Along the creek beds and gullies clumps of brush still burned and smoked. A cloud of ash was beat into the air by the passing hooves, coating all a pasty grey. Everyone, man and beast, was beset by a dry cough.

George pondered the work of fire. It contributed much to the rich, black soil, providing nutrient and space for new growth, but what a fearful experience to be in the midst of it. This whole prairie landscape was one of extremes and chaos: wild fires that burned up everything, locusts that ate nearly everything – both driven by a wind that smelled of great distance – and the land itself, long ago pulverized and scrambled by glacier ice.

The absence of feed made the return journey hard on the animals, and the continual freezing of food tested the endurance and clothing of some of the men. George carried his ink bottle inside his shirt to keep it from freezing. On arrival at Fort Dufferin, he noted how some men's clothes were torn almost beyond recognition, while others' boots had gaping cracks in the soles, and some wore crude suits fashioned out of their remaining blankets. His own were not much better, he realized. A cartoonist would have a heyday with their dilapidated condition.

∞

Before Christmas, George was back in Montreal, unpacking his crates of samples and specimens much as his father had when George was a boy. With the assistance of his father and other specialists from McGill College, George spent the winter identifying species, writing out his field notes, and preparing for the next summer's fieldwork.

The position of Surveyor for western Canada, with the Geological Survey, did not come through that winter, for George or anyone else. Again, George had Dr. Selwyn's assurance that the position would be his once approval was obtained. So George returned to the Boundary Survey the next summer, eager to move west and see the Rocky Mountains.

C-081792

"The last generation of buffalo hunters." Assiniboine Indian Camp, 17 July 1874.

2

1874: Where the Buffalo Roam

Owing to the vastness of the region covered
by the operations of the survey, much of the
period actually spent in the field has been
necessarily employed in more or less ardu-
ous, and often almost continuous travel.

N ext April found George clinging to his wooden
seat as the stagecoach lurched across the bald
prairie of North Dakota towards Fort Dufferin.

Heading north from Moorhead, the driver abandoned the potholed road in favour of the frozen prairie
sod, driving many extra kilometres to skirt round the
ends of flooded coulees and half-frozen swamps.

George rode up on the box beside the driver, looking out over the surging backs of the horses and circle of prairie, leaning into the cold wind and feeling the occasional wave of sweet warm air on his face from the horses working in the traces. "You travel this route frequently?" asked George, mildly concerned about not having seen the road for some time.

"Last time was, oh, two years ago," replied the driver.

"Ah," said George.

"Don't expect it's changed much," said the driver. "Except for the new bridge on the Buffalo. Shouldn't be hard to find."

As darkness approached, George noticed the driver mumbling to himself as he scanned the dimming prairie. After some groping about along the riverbank the new bridge was found, but the approach to it cradled a large, half-thawed snowdrift.

"Hyaaa!" The driver goaded the team forward into the drift. Snow flew as the horses floundered, and in no time the coach was stuck fast in the deep snow. George and the passengers walked the final distance of about one kilometre to the station at Georgetown.

They arrived late at Elm River, found there was no dinner to be had, and pressed on to the next station at Goose River. Dinner was the usual bread, fried fat pork and potatoes, with doughnuts for dessert. Tea was always served, and it was either absolutely tasteless or bitter as gall.

Two days later, and only five kilometres out, the coach dropped with a jolt into a water hole in a coulee. George grabbed his seat to keep from pitching forward as the driver flew off the box, still holding the reins to the horses and front wheels, which had become separated from the coach. All the luggage and cargo had to be unloaded before the coach could be pulled from the coulee and the wheels replaced. Underway again, they came upon a second snow-choked coulee and had to unload the coach a second time and drag it and the luggage across on blankets and buffalo robes.

By the time they reached the third coulee – this one full of water and rotten ice – George noticed that the sips the driver had been taking from the bottle in his coat had taken effect.

"Hah!" said the driver. "We'll do it right this time." He winked at George with a bloodshot eye.

They unhooked the horses and led them across, then began to haul the coach. Midway across, the wheels cut through the ice and the coach was stuck again, another three hours lost.

By the time the coach left the dinner stop, the driver was hopelessly drunk. A few shallow swamps grabbed at the wheels, here and there, until at last the wheels hit a soft spot on one side and sunk to the hubs, very nearly tipping the coach. George and the passengers scrambled out into the ankle-deep water as the driver stumbled and swore at the team. When at last the coach was free and they were moving again, a light became visible in the distance. The driver, though, was too drunk to see it.

"Over there!" shouted George against the wind. "The light is over there!" He grabbed the driver's sleeve and pointed.

"I see it," grumbled the driver, taking up his whip and cracking it over the heads of the horses.

Next morning, George and the passengers were hesitant to board the coach. The driver noticed their reluctance and addressed them with his thumbs hooked in his belt. "You may be expecting an apology from me," he said. "But such is not the case. I have no respect for the man who will not get drunk on a road like this one from time to time. All aboard!"

George and the passengers breathed a collective sigh of relief when, two days later, they arrived in Fort Dufferin in one piece.

∞

As the Survey headed west again, George often rode out of sight of the main pack train as he lost himself in his musings on the landscape. The flat prairie had a circular feel to it, like a giant wheel with George at the hub of it. All lines of diameter ran through him, and all lines of radius began in him. And as he rode, the geometry of radius, diameter, and circumference of horizon was continuously rearranged around him in an endless array of possibilities.

At Porcupine Creek, the survey crew picked up where they had left off, and resumed their measuring, taking bearings and astronomical readings, and constructing boundary markers. The wagon train pushed on across the prairie and through a stony plateau dappled with swamps and ponds and known as the Missouri Coteau.

Just over a hundred kilometres farther west, the country changed yet again as the plateau gave way to a prairie riddled with coulees and tributaries that flowed into wide, deep valleys containing chains of saline lakes.

The clay banks of Frenchman's Creek, or White Mud River, cut a northwest to southeast gash across the 49th parallel and signalled the start of many kilometres of arid cactus prairie. Below the tangle of wild rose and wolf willow abloom at the top of the cutbanks, George scraped up samples of kaolin, a white clay.

∞

The Cypress Hills lay thirty-two kilometres north of the line of the border and were home to the "Big Camp" of the Métis traders and hunters; a place with no law other than that imposed by necessity and general consent. The Hills formed a low plateau, several hundred metres higher than the surrounding prairie. On the northern side the plateau sloped gradually, while the southern flanks were a steep, broken topography, suitable for camping and wintering over. The ground gave up round, flattened stones, from potato to dinner-plate size; cobblestones smoothed by the action of ice, water, and other stones.

As George passed through the Big Camp with his guide, he counted over two hundred teepees, most of them made from buffalo hides, but some from canvas. Each family, he was told, owned at least one Red River cart, which they grouped together with other families to form corrals for their horses. The Big Camp housed about two thousand horses: buffalo runners and packers. The Métis who dwelled in the Cypress Hills were the last generation of buffalo hunters, and each day the men would ride out for the hunt and kill up to six or eight of the beasts. The women would follow behind to butcher the animals, but would take only the best meat from each animal – the tongue and the hump – and leave the rest to rot. The prairie surrounding the Hills was dotted with rotting carcasses and, when the wind was right, the stench of carrion was overpowering.

∽

The first day George saw the Rocky Mountains he felt like a man, lost at sea, who at last sights land. Though the mountains were little more than tiny blue ridges on the skyline, their obvious size and grandeur, even from such a distance, made his heart race. There! There was the land he dreamed of exploring! There were the rocks and uplifts, the splendid cirques and ridges he wanted to scratch and hammer at! From that day on, the expedition could not move westward fast enough to suit him.

At night in his tent, he could not help but imagine himself working for the Geological Survey and exploring the Rockies and beyond, tramping through the mountain ranges of British Columbia to the Pacific Ocean. He hoped his father was right. Dearly, George hoped that a position with the Geological Survey would be awaiting him upon his return to Montreal that winter.

∽

By the end of July the Boundary Commission Survey was camped in the area of Three Buttes, or Sweetgrass

Hills, along the Alberta–Montana border. The Rockies were now a permanent fixture on the western horizon. On a clear day they were beautifully defined and showed great white patches either of snow or some light-coloured rock on their flanks. Each day some new feature – a ridge or snowfield – revealed itself to George, fuelling his anticipation.

As impressive as the distant Rockies were the great buffalo herds that passed through the Sweetgrass Hills that summer. Warned of the approach of a large herd, George and a Métis guide rode to the top of the highest hill, nearly 550 metres above the plains, to witness the sight. As the great brown mass rolled towards them like a giant wave, George could not see the end of it in either direction. The number of animals was beyond his ability to estimate. Thousands. Hundreds of thousands, perhaps.

"They feed many people," remarked the guide. "Métis, Sioux, Assiniboine, Gros Ventre, Blackfeet, Crow, Cree. All live from the buffalo."

There was a story George had heard, and thought vastly over-inflated, about how a party, only two years earlier, had ridden forty kilometres per day for seven days through a single grazing buffalo herd. It might be true, he now realized.

There were few traces of glacial action in the Three Buttes area, but George did find erratics up to a height of 1220 metres, which he considered the highest and most western limit of these rocks. Rocks, he calculated, that had been transported over 1100 kilometres from their place of origin on the Canadian Shield.

The summits of the Three Buttes, George realized, would have been islands in the sea of ice – a chain of islands poking through a hole in the ice sheet, a refuge from the ice which advanced and receded on the shores of the islands like a giant tide, with a cycle measured not in hours or days, but in thousands of years. Buffalo and grizzly bear would have been unlikely neighbours here, seeking shelter in the brush-choked coulees from the daggers of wind.

∞

The Three Buttes area was considered an area of neutral ground for several neighbouring tribes, and when they did meet it was usually in war parties, travelling fast and light. This was confirmed when one of the surveyors returned from his day's work claiming to have found a party of dead Crow warriors.

"Twenty-one of the poor sods," he explained to the gathered crowd. "'Bout ten miles nor'west of us. Just layin' there on the prairie where they died."

"We should go bury them," one man suggested.

"I'll take my camera," George volunteered.

The next day George packed up his camera and extra glass plates and joined the crew on the sixteen-kilometre ride to the scene. They came upon the dead men from over a low rise, the remains intact and in a mummified state from the cold, dry winter.

George set up his tripod and camera and made an exposure of one of the surveyors sitting on the ground amid the corpses. Then, he knelt among the dead and examined their condition. They were remarkably well preserved, with skin tightly stretched over bones that protruded where the wolves and birds had been depredating.

"I heard about this massacre from a man back at the Big Camp," said one of the engineers.

"Who killed them?" someone asked.

"Blackfeet, likely," said the engineer.

George rose and walked to the top of the rise. From that vantage point the story of the fight was plain. The Crows were on foot and their adversaries on horseback. On finding themselves surrounded, the men had dashed up the slope to the empty badger holes, which they hastily scooped out, piling stones around the edges to form shallow rifle pits.

George pictured the Blackfeet as they began to ride round and round at full speed and fire at the dug-in Crows. They used their horses as shields and fired from under or over the neck, offering poor targets.

The fight might have been over in several hours; or, it might have taken several days as the Blackfeet waited for the Crows to grow thirsty, weaken, and die from their wounds. They had no water and could not hide from the sun.

When the Crows were all dead, the Blackfeet had ridden in among the bodies and dismounted, unsheathed their knives and scalped each of the dead men, removing skin and hair from the forehead to the back of the neck. With their rifle butts they smashed in the skinless skulls, then took up their knives again to cut and slash the bodies in all directions. Then, they collected the dead men's weapons, mounted their horses, and rode back to camp, proud of the story they could tell.

∞

"Stampede!"

George looked up from sorting his rocks, saw nothing on the horizon, but was instantly aware of a dull and distant roar.

"Circle the wagons!" someone shouted, and suddenly teamsters were diving for their reins and barking at their teams, as routine calm became chaos.

"Grab your rifle!" the cook shouted at George as he ran past. "Lots of ammo!"

George nodded wordlessly, loaded his rifle, and began jamming bullets into his pockets. The roar in the distance began to grow loud and menacing. He could feel tremors in his leg bones as it travelled through the ground, messenger of the avalanche to come. He joined a group of men behind a row of wagons and found a rest from which to sight his rifle. Behind him, teamsters were fighting to control their teams and bring their wagons into position. Horses and oxen fought the traces as the thunder swelled. George was momentarily fascinated by the trembling in the bones of his entire body.

"Ready to fire!" came the command, and George focused his attention on the brown roiling mass of rising and falling humps now visible in the near distance. He nestled his cheek against the rifle and sighted into the herd, squinting to distinguish one animal from the next.

"Ready!"

George took a deep breath and released it slowly. How could they possibly stop a stampede of such magnitude? Rifle bullets seemed hopelessly inadequate.

From the corner of his eye he spotted movement, heard war whoops and shouts pierce the roar of hooves. A dozen Métis men had broken from the line of wagons on horseback and were galloping straight for the oncoming buffalo.

"Hold fire!" The men watched with disbelief as the Métis charged towards the stampede, screaming war cries and firing rapidly. Several buffalo stumbled and pitched headlong to the ground, impeding those behind them. Miraculously, it seemed, the brown avalanche parted and rumbled past the wagons on both sides with a roar of pounding hooves that was deafening. Though George could see nothing through the choking dust, he kept his rifle aimed into the herd. But there seemed no need, now, to fire a shot.

After five minutes or twenty-five minutes – George could no longer track time – the thunder began to diminish and was punctuated by the cries of Sioux hunters on horseback, as they charged past the wagons with barely a sideways glance, picking off the slower animals.

After the dust had settled, the scene was like the aftermath of a great battle. Fresh buffalo carcasses dotted the prairie as far as George could see. For several days to come, fresh buffalo meat was served with each meal.

With his belly full of buffalo meat, George spent the next days riding through a landscape trampled by thousands of hooves. Fescue grass was crushed into the soil; chokecherry and wolf willow lay smashed and pulverized where the bison had rushed down coulees and creek banks. The devastation was as thorough as that of wildfire or locusts.

George stopped to water himself and his horse at a slough where the edges had been churned to ankle-deep muck. He scooped the water to his mouth, tasted it, and spat in disgust. It tasted of urine. *My God,* he wondered, *how many buffalo have stood in this slough, drinking and excreting?* Even the horse's thirst was shortened by the bad water.

Though he may not have known it, George was witness to the last of the great buffalo herds. By the end of the century they would be hunted to the brink of extinction.

∞

In Blackfoot country, George began to find shallow, sandy pits rimmed with stones, usually near the crest of an isolated hill. He asked one of the Métis men what they were. "Dream beds. Young Blackfeet men make the pits and lie in them until they receive their vision."

"How long does it take?"

"Days. Sometimes a week. Sometimes more."

"What do they do about food and water?" asked George, intrigued.

"Nothing," came the reply. "Lie there until their spirit visits them."

George thought little more of the dream beds until he rode up on one several days later. It was a still, warm day without the usual wind to lean into. He dismounted beside the sandy depression and examined the dream bed, trying to imagine some young warrior lying there until thirst, exposure, and desire brought him the vision he sought, his passage to manhood.

George stepped into the pit and sat down, scanning the empty prairie. His horse grazed contentedly nearby. He lay down, felt the sun-warmed sand through his shirt, stretched out his legs and, staring up at the lid of sky, imagined the warrior. A single locust rubbed its wings together, testing the air, then fell silent.

The silence began to groan as if the whole of earth history was flowing into the present like some giant, slow-moving river, flowing into the moment of George Dawson lying in a shallow bed of sand rimmed with stones on the empty prairie. The air buzzed with the creaking passage of time. Nothing but horizon and sky to distract a man from his vision. Ice, some 1.6 kilometres thick, had once lain over this land. Jurassic, Cretaceous, Tertiary, Pleistocene, Holocene: the periods and epochs advanced on a scale difficult to fathom, while the sun advanced fifteen degrees an hour, three hundred sixty-five degrees a day. How many degrees must the sun advance before a vision came? How many degrees had the sun travelled since the ice had retreated from sight?

George sat up. His horse was grazing a hundred metres away. Otherwise, the prairie was intensely empty. How long had he lain there, in the dream bed?

∞

By the end of August, the Survey had reached the west side of Waterton Lake. George was finally in the Rockies and able to climb peaks and traverse ridges. He made sketches of the vista in all directions, cross-sectional sketches of Kootenai Pass, and sketches of individual mountains with their various types of rock detailed. And he chipped away at the different layers of stone to determine origin, type, and the method of formation of the dramatic eastern slopes.

George watched with a certain sadness on the afternoon when the pillar from the 1861 B.C. survey, atop the ridge west of Waterton Lake, was confirmed as marking the 49th parallel, thus joining the work of the two surveys and completing the Canada–U.S. boundary from Lake of the Woods to the coast of British Columbia.

They had built 388 cairns or pillars and established 40 astronomical stations along the 49th parallel. The 1280 kilometres they had surveyed along the 49th parallel was equal to one-twentieth of the earth's circumference.

By October 11 George was back in Fort Dufferin, having completed an overland journey of 1376 kilometres that season. Again, he returned to Montreal to work up his field notes and prepare his final report. As he sifted through his notes on glaciation, he was forced to conclude, like his father, that the only feasible means of explaining the great drift deposits of Laurentian rocks, some transported hundreds of kilometres, was by large icebergs floating across a great inland sea. Not all the evidence, however, supported this theory. The complete absence of any marine animal remains or fossils was a problem, even if the inland sea had been fresh water.

The publication of his now famous "Report on the Geology and Resources of the Region in the Vicinity of the Forty-Ninth Parallel," brought George immediate attention as a scholar and scientist. The Report ran to 387 pages and contained complete descriptions and analyses of the geology and geography of the prairies, with chapters on the potential for settlement and resource-based industries, as well as appendices containing inventories of plants, animals, fossils, and natural resources.

Though that winter was the end of his work with the Boundary Commission, it was, upon publication of his report, merely the beginning of George Mercer Dawson's career as a remarkable field geologist. But before the accolades could reach him, George received the appointment he'd been waiting for for the past two years. On the first of July, 1875, George Dawson was appointed as geologist with the Geological Survey of Canada (GSC). On the first of August he arrived in British Columbia.

Dawson Family Portrait,
steps of McGill College Art Building, 1865.
George at left, back row.

3

A Boyhood Adventurer

I snoeshoed a little and sailed my boats we then played tag round the College Ohara came in today and they played old man. I had a very bad headache so I did not play but went to bed early.

The two boys hiked up beside the creek that tumbled its way down the side of Mount Royal. George and his new friend, Dan O'Hara, had built a pair of toy canoes that were ready for their first voyage.

"We'll have to portage here," said George, jumping out onto a rock and pointing at a narrow chute.

"And there," said Dan, pointing farther upstream. "How many bundles of furs are we carrying?"

"Two each," said George. "Two hundred pounds. Simon Fraser is steering my canoe," he declared. "On his way to the Pacific." He could easily imagine the famous explorer's journey down the great river named after him.

"Samuel Hearne is in mine," countered Dan. "Going down the Coppermine to see the Eskimos."

"He was searching for copper," George corrected him.

"But he saw them," said Dan.

"Yes, but it wasn't the purpose of his journey. It was just something he did because he had an interest."

They crouched at the pool's edge beneath the canopy of budding maple trees. Birdsong filled the air, urging spring to take hold and warm the world. The boys set their frail birchbark canoes upon the water and eyed them for trim. Each boy had a stick for steering and, one at a time, each gave his own canoe a gentle poke, spinning the light, tiny boats as easily as if a breath of air had moved across the little pool. Back and forth across the pool they coaxed the canoes.

"They look fine," said Dan.

"I'm going to put a couple of stones in mine," said George. "Make it more stable."

"Good idea," said Dan.

They steered the canoes to where the water gathered and poured over a narrow ledge into a second pool. George's canoe teetered briefly at the top, as if hesitating, then dropped lightly into the pool below. Dan's followed close behind.

"Aren't they beauties?" said George, admiring the way the canoes danced and spun on the chuckling stream.

And so, they set out on their journeys: Simon Fraser piloting George's canoe down the mighty Fraser River, and Samuel Hearne navigating the rapids of the Coppermine River in Dan's. The boys leaped from stone to stone, from bank to bank and back again, their boots splashing in the cold water as they guided their canoes through rapid after rapid.

"If you capsize you lose all your gear and have to camp out with nothing," shouted George.

"And if you can't build a fire you freeze to death," added Dan.

By the time they descended the length of the college grounds both boys were wet to the knees, but the voyage had been a success. Though each canoe had capsized once, they had otherwise withstood the fury of the rapids and rocks. With one large pool left to cross, George gave his canoe a firm push, hoping to cross in one go and retrieve it on the other side. But midway across, the canoe stalled and sat becalmed in the pool's middle. George waded in to his knees and reached out with his stick to give a final push, but the stick was too short. He took a step deeper and felt the water's grip rise above his knee. As he placed his foot on the bottom the rock rolled under it and George plunged face first into the cold water.

"George!" Dan shouted.

"Aaargh!" growled George as he dragged himself ashore. "It's cold! Really cold!"

"You have to make a fire!" said Dan. "Or you'll freeze to death."

"Where's my canoe?" George stood on the bank, dripping and shivering.

"You'd better go home," said Dan.

"Just help me get my canoe, first."

By the time George reached home, in the west wing of the college building, he was shivering madly. His teeth chattered like a telegraph key with an urgent message.

"George! What happened to you?!" his mother exclaimed.

"Fell in," he mumbled.

"Here, let's get you into some dry clothes. Anna! Heat some water for your poor brother!"

"Oh, George," said Anna. "What have you done this time?"

"Nothing, sis. Just got wet. But the canoes were splendid!"

∞

George Mercer Dawson was born on the first day of August, 1849, in Pictou, Nova Scotia, along the southern shore of Northumberland Strait. He was the second son of John William and Margaret Dawson, although his older brother, James, had died less than a month before George was born.

They lived with George's grandfather, James Dawson, a stern, Presbyterian Scotsman whose home was run in a strict fashion. Grandma Dawson, who seemed to George forever unhappy and frequently praying, died when he was five. Grandfather Dawson had been unable to attend university due to a childhood attack of smallpox which severely weakened his eyes, but he always urged George and Anna that they should attend when they were of age.

George's father, William, worked in his father's printing and book-selling firm and pursued his geological studies in his spare time, exploring the rock formations along the shores of Northumberland Strait. The year George was born William began lecturing at Dalhousie College and three years later became Nova Scotia's first Superintendent of Education. This job required William to travel to all the school districts, and his return home was always anticipated with great excitement by George and Anna. A passionate geologist, William collected samples everywhere he went, so that when he returned to Pictou his bags were rattling full of fossils and rock samples. It was like Christmas when he opened his bags, and George would help his father name and classify each fossil and record its place of origin. Keen to stay abreast of his local geological studies, William would often take George and Anna into the field on weekends, lecturing them the entire time on the landscape and formations through which they were walking or riding.

∽

When George was six, his father was offered the principalship of McGill College in Montreal, Quebec, and that fall the family moved into the west wing of the Arts Building on the McGill campus. Although in the city, the college grounds were still semi-wooded and grazed by cattle, and the flanks of Mount Royal provided a fine wilderness for the growing Dawson children – George, Anna, and William Jr. – to explore.

On Saturdays, Mother would dress the three children for an outing, and off they would go with their father through the city with its copper-roofed buildings and magnificent cathedrals, down streets lined with maple trees and tall, narrow houses with second-storey staircases. Father would take them to visit some place such as a post office or a printing press, so the children could learn how they operated; or they would set off for some uninhabited place outside the city to collect geological or botanical samples that their father would use to teach them about their new home.

∞

When George was eleven, he was as active and happy and eager to learn about the world as a boy could possibly be, but that winter, pain began to creep into his young body, causing him headaches and keeping him awake at night. By spring he seemed better, and through the next summer he played as hard as he liked, launching toy canoes in the creeks, fishing, exploring Mt. Royal, capturing butterflies, and tending his garden.

But as winter returned, so too did the headaches and pains in his back and neck, and for much of that winter George lay in bed, unable to rise because of the pain and unable to sleep properly. Wild dreams and nightmares visited him at night, making sleep difficult and frightful.

Anna became his nursemaid. She brought him books he requested and spent time by his bed telling him about her days at school with their friends. In the evenings, if George was not in too much pain, his father would tutor him on his school lessons so he wouldn't fall too far behind.

Not one of the local doctors could make a definitive diagnosis. To the horror of his mother and father, the doctors discovered that George's back had become deformed. His spine had grown hunched and twisted inside his small body. They prescribed medicines, such as cod liver oil, milk and lime water, gold salts, arsenic, iodine, mercury bichloride, sarsaparilla, and walnut leaves. None had any effect on George's painful condition.

That summer George was sent to Pictou to live by the sea with his grandfather, with the hope that the sea air would improve his health. But in Pictou he missed Anna's care and attentions, and though his grandfather was caring and affectionate, no one but Anna knew when to bring George just the book he wanted to read. After a summer of continued deterioration for George, his father arranged for a visit to an orthopedic specialist near Boston.

Dr. Buckminster Brown immediately recognized the problem, as he himself was a victim of Pott's disease, the very illness that was causing George's pain and deformity. Pott's disease is tuberculosis of the spine, a slow-acting bacterial disease that causes increasing pain as it softens and collapses the vertebrae. As the vertebrae deteriorate, the spine curves and twists itself out of shape.

George learned that he was facing the threat of paralysis of his lower limbs and, worse, that 30 per cent of people afflicted with Pott's disease died. That he might not walk again, if he should live through it, was a tremendous shock to George and his family. He wondered what kind of God would punish him in this way? And for what reason? He had done no evil that would justify such punishment.

Dr. Brown prescribed medicines for George, and had him placed in a truss to immobilize his upper body and prevent the deformity from becoming any worse. The medicines were bad enough, even when his mother mixed them in applesauce, but the body truss was a humiliation. It made him completely immobile and therefore totally dependent on others, which he hated. Thank God for Anna, who tended to him tirelessly and always cheerfully.

It was like a prison sentence for an active boy like George to be placed in a body truss and confined to bed. No longer could he roam the college grounds or the mountain with his friends; regular school was no longer possible; the pain in his head and back and legs was nearly constant as his body battled the disease.

His back, he now fully understood, was permanently disfigured – he would never grow any taller. He was a hunchback.

∽

George was nothing if not determined, and over the next three years his health improved slowly. The headaches came less often and were not as severe; his strength gradually returned, and he spent more and more time out of bed, moving around the house in a hand-carriage, or wheel-chair, or sitting by a window to watch the world outside, where he longed to be.

He read every book in the house; from his father's own textbook, *Acadian Geology*, published when George was six, to the novels of Charles Dickens and the poems of Lord Byron and Tennyson. His tutor, whom he called "Miss," kept George at his schoolwork so he would not fall behind.

George spent his fourteenth summer with his family and friends at Tadoussac, a resort town along the St. Lawrence River, where he fished and boated and even went on a few hikes. There was so much to see and learn out of doors, and it had all been passing him by while he'd been laid up in bed. His appetite for the natural world had grown insatiable. To be able to explore it again was like being born a second time.

One evening at dinner, William brought up the subject of a new hand-carriage for George. "You've about worn out the one you have now," he said. "There are some very fine hand-carriages available now. What do you think, George?"

"Yes, George," said Mother. "You've been very hard on this one."

"Well," said George, after a moment's thought. "If you can find one suitable for running up and down precipices and through tall woods full of spruce and scrub, then I suppose it might do!"

"Oh, George," said Mother. "Be serious, will you please?"

"I am serious, Mom," he replied. "I have no intention of spending the rest of my life in a hand-carriage. How in the world will I get any work done if I have to wheel myself around in one of these ridiculous contraptions?!"

"You could be a teacher, or a professor," she offered.

"No," he said. "I want to go exploring."

∽

The woods around Tadoussac provided an unending supply of projects for George, from dissecting conger eels to trying to hatch moth eggs to learning about silkworms. He also began to acquire skills that he would put to good use in the years to come. For example, George and Dan O'Hara learned to ride horseback quite by accident, when they borrowed an old mare and buckboard wagon one day to go to the beach.

The beach along the St. Lawrence extended far from shore at low tide, offering all kinds of opportunities to discover new kinds of marine life. Leaving the mare hitched to the wagon and tethered to a tree in the woods, George and Dan set out across the mud flats, each with a bucket.

A light breeze ruffed their hair and the cool marine air was nearly chilly, though the sun shone from a clear sky.

"I like the breeze," said Dan. "Keeps you cool."

"And keeps the blackflies away, too," said George.

By afternoon George and Dan had in their buckets three crabs, a dozen or so periwinkles, and several clams they'd dug with a stick. They trudged across the shrinking flats towards the woods, carrying their shoes.

"Tide comes in fast, doesn't it?" Dan observed.

"Yeah," said George. "You notice it more on a flat beach."

In the shade of the woods the blackflies were waiting. Beating them away with their hands, the boys returned to the horse and buckboard.

"Oh, no!" they chorused, when they saw the broken shaft.

"She must have started kicking to get rid of the flies," said George.

"What'll we do?" said Dan.

George set down his bucket and stroked the mare's forehead. "Poor girl. We'll have to unhitch her and ride home."

"You ever ridden a horse?"

"No."

"Me either."

"Guess we'll learn, won't we?" said George, a twinkle in his eye.

When George reined in the mare in front of his family's summer cottage his mother appeared at the door in an instant.

"Where on earth have you been? What happened to the buckboard? George, you don't know how to ride. Get down off that animal at once!"

"Mom. It's okay."

George's father came to the door to appraise the scene, his hands tucked into his vest pockets. "Well, well," he said. "It looks like George has taken up horseback riding."

"William, help him down," said Mother.

"I'm afraid to ask where the wagon is. Did you enjoy riding?"

"Oh, yes, Papa," said George, "It's wonderful!"

"Perhaps we should arrange some riding lessons," said William as he lowered the boys to the ground.

∽

Photography had also become one of George's hobbies, and his father gave him his own camera and kept him supplied with glass plate negatives and chemicals for developing his pictures. Taking photographs was much quicker than making sketches, he realized, and more accurate.

∽

The summer George was seventeen, his father presented him with a small rifle and a handful of rounds.

"Take this out back and practise your aim. And remember our earlier lessons."

"Yes, sir. Thank you," said George, accepting the rifle with the reverence deserving of a great prize. It *was* a great prize. George had wanted a rifle of his own ever since the first lesson his father had given him.

That afternoon George went to the back of the garden and set up a small wood box at the end of a row of peas and began shooting. His father watched for several minutes; then, satisfied that George wasn't going to harm himself or anyone else, he retired to his study.

At first, George fired from close range, hitting the box with nearly every shot. Then, he began to move away from it, increasing the distance of each shot. Most shots hit the box, but some of the rounds missed and went cutting through the pea plants like a knife.

By the time George finished all the rounds the wood box was riddled with holes, and the peas close by were chopped and cut and ready to go into a salad.

∞

For seven long years George was a semi-invalid. At times he was able to live a normal life, to get out of bed in the morning and eat breakfast with his family before walking to school, while at other times the terrible headaches returned and confined him to his hand-carriage or bed for days at a time. He detested the idleness forced upon him, and when possible he kept busy reading, sketching, cleaning, sorting, and labelling specimens for his father, and writing stories and poems of his own.

∞

The summer he turned nineteen, George registered for classes in English, Chemistry, and Geology at McGill College. By the following spring he had made up his mind to follow his father's path and study geology. The Royal School of Mines in London, England, offered a three-year program in geology and mining, and George wanted to attend. After some discussion, his father and mother agreed that George was well enough to go. The British school acknowledged and complimented William Dawson's stature as a geologist by granting an exemption of tuition fees for his son, George.

"The high country, where marmots whistled from the rocks and patches of snow still lay on the ground."
Northeast Peaks, Ilgachuz Range, 22 July 1876. Exposure: small stop. 2.5 min.

4

1875: Bushwhacking and Brulé

Two or three years in this dead & alive place would cause anyone to lose all idea of the value of time, & quite spoil one for life anywhere else....

One might as well be in Timbuctoo – or Victoria.

U pon receiving his appointment to the Geological Survey of Canada (GSC), George travelled by railway to San Francisco, then boarded a steamer for the 1150-kilometre journey to Victoria, B.C. The voyage was, quite literally, a voyage of discovery, as neither the ship herself nor any of the officers had been there before.

Out of sight of land, George studied the deep blue of the rolling seascape and was struck by its similarity to the prairies. The territory of British Columbia must, in its own way, be as huge and unknown as the prairie had been that first summer. Very little surveying had been done in B.C., with the exception of Mr. Selwyn's initial survey of 1871 and Mr. Richardson's surveys during the past four summers to examine the coal deposits on Vancouver Island. George's job was to explore the province and determine the fundamental geological composition of Canada's westernmost province, and to assess agricultural and mineral resource potential.

For two days, the steamer ploughed through a vast expanse of jellyfish, which looked, from up on deck, like purple sail jellyfish, or by-the-wind sailors. George desperately wanted a sample to confirm their identity.

Casting his eyes about the deck he spotted lengths of rope hanging in coils in several locations. He walked forward as far as the wheelhouse, where he spotted a bucket belonging to the cabin boy. After a quick check to see that nobody was watching, George casually picked up the bucket and returned to the rear promenade deck.

He was down on one knee, fastening the line to the bucket with a series of tightly cinched knots, when a voice interrupted him.

"You're going to need some help with that, my friend."

George looked up to see a man in a dark grey suit and fedora standing over him with his suit jacket pulled back and his hands in his trousers pockets.

"You may be right," said George, as he stood and tested his knots with a tug of the rope. "If you would kindly secure this end," he said, indicating with his foot. "And begin pulling when I say, then it should work."

"What do you think they are?" the man asked.

"*Velella velella*," said George, giving the Latin name. "By-the-wind sailors."

By now a small crowd had gathered along the rail to watch the proceedings, and to speculate on the chances of success. George paid out the line hand over hand. The bucket bounced off the water, and he realized just how fast the steamer was travelling. This was going to be harder than he thought.

He let the bucket drag along the surface and flicked the rope to get the bucket to tip and fill with water, but it just skipped along on its side.

"Harder than it looks," said an observer.

George hoisted the empty bucket and set it on the deck at his feet. "We're travelling too fast," he said, discouraged. "The bucket's too light."

A while later, George watched as one of the other passengers picked up the bucket, lowered it over the rail, and jigged it a couple of times. Suddenly, the rope was running out rapidly, burning the man's bare hands. He dropped the rope and thrust his hands into his armpits, and in seconds the rope had run out over the rail and disappeared in the ship's wake. Not ten minutes later the cabin boy came around asking if anyone had seen his bucket.

∽

After being forced to lay over in Victoria for a week, George finally departed on the 13th of August. Past the row of salmon canneries and stands of giant spruce and cedar trees of New Westminster, the steamer *Royal City* carried him as far as Yale, where he arrived in time to board a stagecoach bound immediately for Lytton, Ashcroft Manor, Clinton, and Soda Creek.

The ride was anything but comfortable, as the coach bounced and careened its way up the road carved into the walls of the Fraser Canyon. George rode up on the box beside the driver, the better to survey the passing terrain.

The road dipped and climbed its way up the side of the canyon, sometimes close to the water and other times climbing up to 245 metres above the river to overcome bluffs and cliffs. Wherever the land was level enough for ploughing, small farms were located. In places where the road ran close to the river, George could see Indians using long-handled dip nets to fish salmon from rickety-looking platforms built out on the rocks beside the rapids.

All along the road, George compared his own geological observations with those of the GSC's director, A.R.C. Selwyn, who had conducted a survey of the canyon four years earlier. Too busy with his observations to notice the bone-rattling ride, George was much impressed by the dryness of the area, and the trees and birds: lodgepole and ponderosa pine, sagebrush, magpies, Steller's jays, and Clark's nutcrackers.

At Soda Creek, George found that only partial arrangements had been made for him. There were three pack animals and two riding horses, as requested, but no cook and no guide. The steamer crossed the Fraser the next morning and not again until the following Sunday, so George decided to travel upriver to Quesnel to see about completing his outfitting there.

∞

When George returned to Soda Creek, his party consisted of himself; Reeves, the cook; two Interior Salish brothers, Jimmy and Tommy; and three packhorses and two riding horses.

The next morning, George and Reeves arose at 4:00 a.m. and crossed the Fraser by canoe to where Jimmy and Tommy had spent the night with the horses and supplies. They loaded the pack animals and set off up the side of the river valley, southward, as dawn flushed into full daylight. Near the rim of the valley, and not an hour into the summer's expedition, George heard a violent crashing in the brush behind him, followed by the shouts of Jimmy and Tommy. He turned to see one of the packhorses rolling and crashing down the bank through sapling spruce trees.

"Quick!" he yelled. "Get her!" He dismounted and tethered his own horse to a tree, then went scrambling down the bank to the fallen mare. Jimmy was already with her, coaxing her back to her feet. "She's okay," he said. "Nothing serious."

"We'll see," said George. "Lead her back up to the trail."

As they readjusted the load, George could see the mare was favouring one leg.

"We'll have to lighten her up," he said. "Here, transfer some of this to Reeves's horse and some to mine." He looked up at the sky as raindrops began to patter down. The rain released a wonderful resiny smell in the dry pine forest, but that momentary pleasure was soon forgotten as the rain grew to a steady downpour that lasted the rest of the morning.

Along the benchland high above the river they rode through open timber and patches of luxuriant, tall grass. George jotted in his notebook the Latin names of plants: *pinus contorta* (lodgepole pine), *populus tremuloides* (trembling aspen), *pseudotsuga menziesii* (Douglas fir), *Sheperdia Canadensis* (soapberry), and many others.

They turned west to Riske Creek and Alexis Creek, along the Chilcotin River, and after three days along the north bank of the Chilcotin, they forded the river and travelled south to Tatlayoko Lake, deep in the Coast Range mountains.

West of Alexis Creek, where they'd seen a number of shanties and gardens of potatoes, carrots, and turnips, George's party stopped to camp near a large band of Chilcotins. In broken English and Chinook Jargon, a trade language used among northwestern tribes, they told George they were a band of Chief Alexis's men, on their way to the Fraser River to fish.

They had just that day caught the first salmon of the season, and they presented it to George and his men, a great honour. In return, George gave the Chilcotin men tobacco and pork. After supper he offered them all a cup of tea and the remainder of their grouse stew.

"Good, this tea," nodded their spokesman, after saying grace. None of the men, however, touched the stew pot.

"You don't want the stew?" asked George.

"Friday," said the Chilcotin. "Friday no meat."

Then it struck George. A Catholic priest had instructed them not to eat meat on Fridays as part of their Christian faith. He smiled at the incongruity of it. "Tomorrow, then? You'll eat it tomorrow?"

The Chilcotin smiled and nodded in reply, and true to his word, the stew pot was returned in the morning after breakfast, empty and wiped clean. Along with the pot was another salmon and a trout, which George traded for with tobacco.

As they rode on to the crossing of the Chilcotin River, George considered his party: Reeves provided camp fare that was better than many meals George had been served in "civilized" eateries. He was a man of about fifty, with a strong Cockney accent, who had run an eating-house in London, England, until rheumatic gout drove him out of business. On the advice of his doctor he went to visit his sister on Vancouver Island and was passing the time by hiring on with various survey parties as a cook. He was "seeing the country," he said, and planning to return to London after a couple of years in Canada.

Tommy and Jimmy, the Salish brothers, were working out satisfactorily, though Jimmy, the elder, seemed to consider that Tommy, the younger, should do the hardest work. They seemed to get along okay, and that was fine with George. His only concern was their capacity for food, which was about double that of himself or Reeves.

They were camped along Tatlayoko Lake, near where the Waddington party was massacred in 1864 as they sought a route to the coast for the railway. The lake was rimmed with steep, snow-capped peaks, and in the mountains at the south end of the lake was a glacier, the first George had seen, a massive remnant from an age when ice was the dominant feature of the land-scape. Hiking around the rocky hillsides, George found the characteristic scratch marks, or striations, left on the bedrock by the glacier. He lay flat on his belly on the mountain to sight his compass along the striations, and recorded the direction of the flow of ice. If the drift-ice hypothesis, which he and his father subscribed to, were to hold true here, the land would have to have been submerged more than a thousand metres to allow for drifting icebergs to leave such marks. The thickness and weight of the ice would have been incredible. The alternative theory – glaciation – was perhaps a better fit in this instance, but George was not yet convinced enough to abandon the drift-ice hypothesis.

One morning, by 7:00 a.m., George, Tommy, and Jimmy rode along the lakeshore towards a creek reported as a good fossil site. Grouse hooted in the woods at their passing, and along the lakeshore mallard ducks turned and dipped in the shallows. At the mouth of Fossil Creek, George reined in his horse.

"Jimmy, you stay here with the horses. Tommy and I will go up the creek for samples."

"Okay, Doctor George."

"We'll be back for lunch."

George and Tommy set out along the bank of the creek, up a steep-sided hill. Up and up they climbed over small cascades and tangles of logs, until they were about three hundred metres above the lake. Just a little higher, George spotted a rock outcrop that looked like a good spot to begin.

He chipped at the hard blue sandstone with his hammer, striking this way and that until he managed to free a chunk. He bent to retrieve it. "Here," he said, showing Tommy. "See this?" With his fingertip he traced the outline of a snail shell in the stone.

Tommy nodded. "*Exogyra*," said George. "A kind of marine snail. Very old."

Tommy looked puzzled.

"A long, long time ago," George began. "This was under water. The sea covered all the land."

Tommy's face brightened. "Okay," he said. "The big flood. Sure. I know that story."

The flood story, which Tommy had probably heard from one of the priests travelling through the country, wasn't what George had in mind. Rather, he was trying to imagine the story of the era when these snails had been alive beneath the ocean's surface. Massive changes had caused them to be perfectly preserved within the sandstone rocks of a place that now stood over 1200 metres above the sea. After George had obtained a couple of decent samples he was content to return to camp.

∞

The rough country and the less than perfect packing by Tommy and Jimmy took its toll on the packhorses, and within two weeks all had sores from their pack frames. Two of the horses had taken tumbles while fully loaded. After three days of tending to gear, changing the paper in the plant presses, and exploring everything of interest in the valley, George's patience had grown thin.

He was tired of waiting for the mules that were supposed to be en route from Quesnel. George arranged with Mr. Cambie, of the Canadian Pacific Railway (CPR) survey, to swap for two of his horses so that he might be underway again. Mr. Cambie kindly offered George his own riding horse so he could begin the journey north to the Blackwater River and Fort George and, though still one horse short, George gratefully accepted and set out the next day, with two pack-horses carrying the loads of three.

Travel was difficult through areas of great windfall and burned patches of fallen timber, called brulé. At times, so tangled and impassable were the stretches of brulé, the countryside reminded George of a giant game of pick-up sticks. Poor feed led the horses to wander at night, and poor packing with oversized loads caused them to grow sore after only three days on the trail.

George sat by the dying fire after a day that had begun with a search for the horses, followed by a loss of two hours due to taking a wrong trail, followed by one of the packhorses rolling down a bank, and ending with Reeves falling from his horse, saddle and all, while climbing a steep, rocky trail. They had been joined that afternoon by Charley, a Chilcotin man, who was on his way to Fort George. He sat with George, staring vacantly into the fire.

"Doctor George," said Charley, as George closed his notebook. "I have a story for you."

"A story?" replied George, who was by now functional in Chinook and quickly learning the Chilcotin language. "Do tell me, Charley."

"Well," said Charley, poking the fire with a stick. "Long ago, three generations at least, maybe more, some bad siwashes from the salt chuck came and camped on top of a cliff along the Chilcotin River. From there, they watched the Chilcotin people pass by, back and forth, and they would shoot them with arrows as they passed by below. No one knew how it was done. No one ever saw anybody shooting the arrows.

"At last a Chilcotin siwash who was passing by one night saw a fire up on the cliff, and he went and told the others. They decided to send up one man while the bad Indians were sleeping and see how many there were. Then, the next day, the others surrounded the cliff and sent a klootchman, a woman, to walk along the valley below to attract the attention of the bad siwashes.

"When the siwashes looked over to see the woman walking below, the Chilcotins came up behind them and killed them all, except for one, who was the medicine man. The medicine man jumped off the cliff into the air and flew away. All same wind Klattawa, all same chicken. He flew away on the wind like a bird."

"He flew?" asked George, laying open his notebook on his legs.

"Yeah," said Charley.

"He must have been a very powerful medicine man," said George.

"Oh, very powerful," said Charley. "All same wind Klattawa. All same chicken."

∞

It was mid-October before George made it to Fort George. Autumn was well advanced, with geese flying in great wedges down the sky and a snowfall that had covered the ground for several days.

Mr. Selwyn, George, Reeves, and Selwyn's assistant set out down the Fraser River to Quesnel, in a dugout canoe and rowboat. None of the men had been down the river before, but they had been given directions for navigating the two canyons.

The morning they set out was foggy and cold, with a layer of cloud lying low over the river. The wooden clunk of paddles on gunwales and the creak of straining oars were amplified by the closeness of the sky and the stillness of early morning. George stood on the gravel shore and watched a line of bubbles rise to the surface and drift apart lazily. Here, the river seemed asleep, but he knew it would not remain so for long. The canyons were a concern, though at this low water level they had been assured the rapids would not be too rough for their boats.

"Ready, then, George?" said Mr. Selwyn, as he stepped into the rowboat and took up the oars. George lowered himself into the canoe and pushed the shore away with his paddle. In minutes they were out of sight of the fort.

The canyons proved to be uneventful, but farther south, where they ran a heavy rapid above the mouth of the Blackwater River, the wind came upstream very strong, raising a short, heavy sea. The air was full of flying spray as the tops of the waves were blown off and flung upstream. George tried to cross the river to a wider channel, but the wind prevented him from doing so, and he drifted down a channel with a boisterous current and waves with their tops flying off and into the faces of George and Reeves. They manoeuvred the canoe through the rapid, getting soaked in the process, and pulled into a pool of calm water at the foot of a cliff.

Above their heads the wind screamed through the pines, and they were forced to dig in their paddles to stay in the eddy. At a crashing sound above them, George looked up to see a large ponderosa pine sliding down the high bank.

"Paddle!" he shouted. "Paddle hard!" The loaded canoe responded slowly to their frantic pulls, and George looked up, fearing the worst. He gave a huge sigh as the tree came to a stop up against several smaller pines.

"That one would have finished us," said Reeves.

"Let's get out of here or it may yet," said George.

∞

A week later George and Mr. Selwyn bounced and swayed with the stagecoach as it careened down the road beside the Thompson River. George and Selwyn rode, as usual, up on the box with the driver, comparing their observations of the summer and the passing scenery. By the end of October the days were growing short, which meant they must travel the last couple of hours to Lytton in the dark.

This was no problem until one of the wheels ran off the edge of the road and onto the sloping bank. Immediately, the coach tilted dangerously and George and Mr. Selwyn found themselves falling down the dark bank towards the river below. The coach rode on two wheels for a moment, then crashed down on its side.

An eerie silence followed, broken only by the horses' nervous stamps and snorts. The driver had managed to hold onto the reins, fortunately, and George heard him talking to the animals as he picked himself up and punched his hat back into shape.

"Get us out!" someone shouted from inside the coach.

"Anybody hurt?" called the driver.

"No! Let us out!"

"Selwyn! Dawson!" the driver called. "You two okay?"

"We're fine!" George shouted as he scrambled up the bank, Mr. Selwyn at his elbow. While the driver worked to detach the horses from their harnesses, George and Mr. Selwyn helped the passengers out of the overturned coach.

The last person to climb from the stage was Mr. Walkem, the premier of the province, who had landed at the bottom of the heap of bodies inside. He now emerged, sputtering and speechless, but unhurt. "Good Lord!" he blurted. "Now I know how a staircase must feel!"

They hitched the horses and pulled the coach upright, loaded all the passengers again, and carried on to Lytton, shaken and a bit bruised, but otherwise unhurt.

George spent the winter of 1875-76 in Victoria, writing up his summer report and collecting specimens for the Centennial Exhibition in Philadelphia the next summer. He worked with Dr. Tolmie to compile vocabularies of different dialects and languages of the First Nations of British Columbia. George socialized only when it seemed necessary to do so, as he preferred to spend his evenings at home, reading.

"The meeting and trading place for coastal and Interior natives." Gatcho Lake and Culla-Culla House with Dawson's camp, 25 July 1876. Exposure: small stop. 2.5 min.

PA-052730

5

1876: Monsters, Tricksters, and Ice

...Where days are
long & woods are dark,
or crowded thick twixt lichened stones
where some old glacier laid his bones.

On the 19th of May 1876, George left Victoria for the Interior again, via the Fraser Canyon. At Soda Creek, he boarded the steamer *Skana* for the trip up to Quesnel. An early spring runoff meant the Fraser was already in flood, and the *Skana* struggled to churn its way upriver. Locals told George this was the highest water they'd ever seen.

After several days of preparation in Quesnel, George and Dan McMillan, an engineer for the CPR, crossed the Fraser and began riding west. Dawson's assistant, Amos Bowman, would follow a few days later with a CPR pack train and their supplies, and meet George along the Blackwater River. The objective for the summer was to survey the territory east of the head of the Dean Channel and Gardner Channel, a vast area also being surveyed by the CPR.

Within a couple of days they were again having to chop their way through miles of windfall and brulé, negotiate swamps, and squeeze through thickets of brush thick enough to scrub a man off his horse. But all was not bleak, as George also found places where many plants were in bloom, such as pin cherries, gooseberries, currants, bunchberries, and saskatoons.

At a camp beside the Euchiniko River, George discovered a small nest attached to a sprig of spruce and overhanging the water. As he approached slowly, he saw a tiny hummingbird sitting on the nest. Her back was green and iridescent, and her throat was white with a series of tiny, rust-coloured spots.

George spoke very softly as he moved to within about a metre of the nest. The nest itself, he could see, was built from some sort of downy material and covered outside with white lichen and black moss. It was amazing that something so tiny and fragile-looking was able to survive a hard country such as this, where men, horses, and mules were tested daily.

A couple of days farther up the Euchiniko River, George and McMillan were drying out beside a large fire when they heard a shot. A minute later they spotted a man moving through the trees towards them. "He's alone," said McMillan.

"No," said George. "He has a child with him. Behind him."

The native man who approached the fire was carrying a huge bundle of furs on his back and was accompanied by a boy of ten or twelve.

"Heard your horse's bell and thought you were Klusklus Indians come over," he said, using a mix of English and Chinook. "Made me angry until I saw you weren't them."

"Where're you headed?" George asked.

"Quesnel," said the man. "Taking my winter furs to sell."

"Dan, put the kettle on for tea," said George as the man lowered his pack against a tree. The boy held his hands out to the fire and drew close to the burning logs. "Where do you live?"

"Up the river," replied the man. "Me and the boy."

"Where are the rest of your people?"

"My father died when I was young, and then my mother and all the rest of them. Twelve, thirteen years ago. Smallpox. One after another I bury them all, but never get sick myself. Connoway mamlouse, nika one stop! Hi-yu sick tum-tum, hi-yu cly. All dead, I one stop! Much grief, much lamenting. Before smallpox good trails in all directions. Now, few trails, much blowdown. Travel is hard."

"Indeed it is," George agreed. "Especially with animals."

"Faster to walk," said the native man.

The boy pulled a bone implement from his shirt and began to work the edge of it on a piece of pine lying by the fire. George watched, fascinated, as the boy made an incision in the bark lengthways and peeled open the bark. With the other end of the bone tool, which was shaped something like a shoehorn, he scraped off the cambium layer and ate it.

The men squatted by the fire and sipped their tea.

"Are you in a hurry to get to Quesnel?" George asked.

"Oh yes. A great hurry."

"I'm looking for a guide," said George. "To take us over to the Blackwater River."

"Klusklus country," said the man. "The way is not clear."

"So I fear," said George. "Would you be interested in working for us?"

"Have to get to Quesnel quick. Sell the furs."

"I'll pay you $1.50 a day."

The man slurped his tea noisily, then said something in his own tongue to the boy, who raised his eyes but said nothing.

"I'll tell you in the morning," he said to George.

Next morning George found the native man hoisting his bundle of furs into a tree. "Ready to go," he said when he saw George. "Just for two days."

"Excellent," said George.

They continued up the Euchiniko River valley under grey skies and showers. Through windfall and brulé they chopped and squeezed and coaxed the horses and pack mules. The boy rode behind McMillan, while his father led the way on foot through the tangled woods.

Just before noon, they came up against a large log, too big to chop through. The only way over was to have the animals jump. It was not a high jump, less than a metre, but it was on a steep hillside, which increased the difficulty.

The first two mules went over with no problem, but the third mule balked with its hooves on top of the log.

"Come on, girl! Hup!" McMillan gave her a slap on the rump. "Come on! Git!"

"Back her off of there," said George. "She can't jump it from there."

McMillan took hold of the halter and pulled back. The mule tossed her head and reared, lost her footing, and fell backwards. The men watched in disbelief as the mule tumbled end over end several times before being stopped by a pile of windfall.

By the time they reached her, the mule was back on her feet, stomping and blowing.

"See if she can make it back up here," George shouted down. He checked the load in his rifle, prepared to put her down. No animal could be expected to survive a fall like that.

"She's okay," said McMillan, leading her up the bank. "Not favouring anything."

"We'll keep a close eye on her," said George. "See she doesn't go lame on us. A fall like that should've broken a leg at least."

After showing them the place where the trail led south to the Blackwater, the native man and his boy parted ways with George and McMillan. Through another day of steady rain they rode down onto the Blackwater, passing through swamps and deep, mossy bogs. At camp, they built three large fires and pitched their tents in the middle to dry out.

Several days later Bowman arrived with the CPR pack train of sixty animals. He and George overhauled their gear and resupplied, working under clouds of mosquitoes and black flies, then continued west to the headwaters of the Dean River.

∞

At Gatcho Lake they camped near the Culla-Culla House, a West Coast style, post and beam longhouse, with a large crow, carved and painted black, adorning the gable above the doorway. On the end of the building, framing the door, were painted in red two large figures resembling bears. Built by the Bella Bellas, the house was used as meeting and trading place for the coastal and Interior Carrier tribes.

Near the Culla-Culla House stood two poles, which, George learned, held the remains of two people. The Carrier convention was to cremate the body and then place the remains in a box in an alcove on top of the pole, which displayed the family crest on the front.

∞

Travel farther down the Dean River to the Salmon House was slow. The river had flooded and made it necessary to cut a new trail on higher ground. The high water had washed out the foot bridges over the Dean, below Salmon House, and delayed the CPR party ascending the valley from the ocean. As George and Amos Bowman waited, they explored much of the area and collected plant samples and searched for fossils.

While waiting for dinner to cook, George took up his hammer and chipped at a small outcropping of bluish, volcanic rock, more out of idle curiosity than anything else. A palm-sized chunk broke free and George examined the newly revealed surface.

"Amos!" he shouted. "A *Belemnite!*"

"A what?" Bowman set down the cooking pot.

"*Belemnite*," George repeated. "A cephalopod mollusc. Look! Do you see the chambered shell there? I'll bet this is the same horizon as those I found at Tatlayoko Lake last summer, which would date them as Cretaceous."

"I would never have expected to find organic remains in volcanic material," said Bowman.

"Nor I," said George. "Nor I."

∽

As they traversed the area, they met native families with all their belongings packed in wooden boxes on their backs, and children and dogs swarming about their feet, on their way to the Salmon House for the fishery.

George stopped to talk to one very old-looking man.

"I remember," said the old man. "When the first white men came here. Four white men came walking from the east. One of them carried a gun, which we had never seen before. They got two Indians from Gatcho Lake to go with them down the Bella Coola Trail to the sea. The Indians came back but never the whites."

"How long ago was this?" asked George, fascinated.

"Many, many years. I was just a boy. Thought the white men were ghosts. Scared me at first."

Could they have been, George wondered, *Alexander Mackenzie and his party, in 1793?* He did the math quickly in his head. *If, say, the man had been a boy of five when it happened, that would make him eighty-three now.* It was possible. He had read Mackenzie's account of the journey, and the explorer had described the Culla-Culla House at Gatcho Lake in his report.

Along the trail taking them back to the Salmon House, George noticed three stone cairns beside the trail, each covered with a pile of small sticks.

"What are those cairns and sticks for?" he asked Jimmy, who had stopped to add a stick of his own to the pile.

"If you pass here," said Jimmy. "And don't potlatch a stick to each cairn, we believe you will soon die. Tenas men are buried nearby."

"Tenas men?" George had not heard of them. "Who are they?"

"Tenas men are a race of short men, dwarves. They used to steal our klootchmen, our women. Some of these men are buried near the Salmon House."

Upon arriving at the Salmon House, Jimmy pointed out to George a circular hollow nearby. "This is where the Tenas men camped and slept," he said. "They made a fire in the centre there."

"Where are these men buried?" George asked, curious to see their burial sites.

Jimmy waved an arm in a vague sort of way, indicating the surrounding forest. "Don't know, exactly," he said. "Somewhere around here."

"Have you ever seen one of them?"

"No. I just heard the stories."

From the Dean River, they travelled south and east to Anahim Lake to check out reports of exposed lignite, a poor quality coal. They climbed steadily, and soon travelled through high country where marmots whistled from the rocks as they passed, and patches of snow remained on the ground. The exposed rock gave George an opportunity to look for glacial striations, which he found at 1120 metres and which ran northwest to southeast. As George lay on his belly and sighted the direction with his compass, he might well have entertained the image of a huge ice mass creeping southward on its belly, carving its initials into the sides of mountains.

Fording the swollen waters of Cheddakulk Creek, Frank the mule lost his footing and began to wash downstream. As he bumped over rocks and tried to regain his feet, the men could only stand helplessly watching and hoping he would not wash over the waterfall below.

Their sighs of relief could almost be heard above the roar of the creek as Frank suddenly stopped midstream against a boulder. To a man, they cheered as Frank struggled to his feet and staggered to the bank.

George was amazed to find Frank had suffered only a few minor bruises and cuts. They checked his packs and found the sugar, tea, beans, rice, and oatmeal thoroughly soaked. George's camera, which had been on the upper side and well wrapped in blankets inside its case, was almost completely dry, and undamaged.

∞

The party travelled steadily northeast towards the Nechako River. Each morning the axemen would leave camp first to chop out windfall and locate the trail, and an hour later George would follow with the pack animals.

Along the Nechako River the travelling became easier. Raspberries, strawberries, and saskatoon berries were ripe, and good feed was abundant. When they finally intersected the Telegraph Trail, George cut off a standing tree at 1.2 metres, squared and marked it: "Pioneer Trail & blazed line to Nechako R, Gatcho L. & Dean R. To Hunter's Camp, (6 miles N of Gatcho L), 100 miles. G.M. Dawson A. Bowman. Geo Survey Canada, Aug. 31, 1876."

Horses, mules, and men had all been pushed to their physical limits by the passage, so they rested at Fort Fraser for a week.

∞

On the 7th of September, George and Amos Bowman set out for Francois Lake in three canoes with four native men, including the two from Fraser Lake, Jason and Charley.

George chatted with the men using bits of English, Chinook, Canadian French, and their own languages. Working in this unusual mixture of languages, he was able to learn the names native people used for the several constellations. The Milky Way was called Ya-ka-tsool-k, which meant "snow-shoe track," and the Aurora Borealis was known as Ni-ha-pa-tun-ut, or "the fire or light coming immediately before the snow."

As they paddled along the shores of Francois Lake, George noticed Charley and Jason had grown very quiet and were looking about nervously.

"Is anything wrong?" George asked.

"Shhh. Klug-us-cho," said Charley.

"Big snake lives here," added Jason. "You bet. Under the water, in the mud."

"Paddle quietly," said Charley.

By the fire that night, George asked Charley to tell him the story about the big snake.

"Oh," Charley began. "Sometimes, when he gets hungry, he listens for us Indians in our canoes. When he hears us he comes rushing up from the bottom and makes a great commotion in the water. We land and run into the forest to hide. You bet."

"Does he come ashore?"

"No, but always, you want to go back and see what he looks like. Finally, you go too close and get caught by his big mouth. When he catches a man in his mouth he tosses him up into the air, over and over, until the man falls head first into his open jaws and is swallowed. You bet."

As George set the fishing net that night, he paused to watch the surface of the water. The lake was perfectly still and dark, reflecting nothing. *How*, he wondered, *did monster stories get started? Were they purely for entertainment, or did they have some other, instructive purpose?* He tossed out the net, watched it ripple the surface and disappear. *Maybe I'll catch the big snake tonight,* he smiled to himself.

In the morning there was no monster in his net, just a nine-pound lake trout, which made a fine breakfast.

∞

On Fraser Lake, they camped near a large, flat, glaciated rock that sloped gradually up from the water. The native people called it *Te-tinn-a-na-nes-tis*.

The striations on this rock, in combination with others George had observed, were yet further evidence to support the glacial theory: great sheets of ice creeping down the valleys, under immense pressure, and being forced out through the gaps along the sides of the basins. The drift-ice theory was beginning to show weaknesses in the face of his field findings. After dinner, George explained how he believed the large, flat rock at the lake's edge had been shaped.

"No," Charley said quietly, when George had finished. "That is not how the rock got its shape."

George sensed a story about to emerge, and said nothing as he settled back on his heels. Charley began to tell one of the stories of Us-tas, an ancient chief, medicine man, and trickster.

"One of the men here had a headpiece like a swan," said Charley. "So he could hide in the water and wade among them to catch them by tying a rope to their legs. Us-tas saw this and said: 'That is nothing. Can you catch only one swan? Let me try.' He put on the swan headgear, tied a rope around his waist, and waded into the lake. Pretty soon he caught five swans and tied them all to his waist. Then he threw off the headpiece and frightened the swans. They all flew up into the air, carrying Us-tas with them, and travelled away to the sea.

"Next spring the swans returned, with Us-tas still dangling by his rope beneath them. As he passed over Fraser Lake, Us-tas remembered his knife and cut the rope. Down he fell onto the rock, which was only mud then, and he sank below the surface. You bet.

"Well, one of his friends rubbed the surface of the mud with grease and went away. Lynx came along and scraped away the mud and grease, scratching as it is now, until the face and eyes of Us-tas showed. Then Crow came along and picked out the eyes and carried them off. Soon after, Us-tas got up and walked into the forest, knocking into trees, for he could not see.

"The Indians laughed at him and said: 'Us-tas is blind at last.' Us-tas said: 'Wait, I know what I am about. I am not blind, but looking for something.' He heard singing in a large house and went to the house and an old woman told him: 'We are celebrating, for we have got a man's eyes from the crow.' Us-tas said: 'Let me look at one.' They gave him one, but he said: 'This is only one,' and held out his hand for the other. Having them both he pushed them into his eye holes and ran away fast, seeing again."

"So the lynx made the scratches in the rock," said George.

"You bet," Charley replied. "Before that it was just mud."

"Do you know any stories that tell about ice in the valley?"

"No," said Charley. "No stories like that around here."

∞

The journey back to the coast took them down the river to Fort George and Quesnel, then overland to Yale, and by steamer back to Victoria. On the 20th of November, George left for San Francisco aboard the steamer *Dakota,* and then rode the train east to arrive in Montreal on December 4, 1876.

6

1877: Salmon and Camels

I feel already as if I had been journeying forty years in the wilderness since leaving Montreal, but hope to enter into the Promised Land via New Westminster about next Friday.

Along the shore of Okanagan Lake, Johnny pointed out a bird singing in a nearby tree. "Wa-wa all same siwash la langue," he said: The bird speaks the dialect spoken by the Indians in each place.

"How does it know what dialect to use?" asked George.

"Just knows," said Johnny. "It listens to the people then talks the same way."

"The ease with which great distances are seen in this country."
Bluffs of white silt, lower end of Okanagan Lake, 1 July 1877.
Exposure: stop no. 5. 7 min.

"And what does it say?"

"Do not steal! Do not steal!" mimicked Johnny.

"A moral bird," George observed.

At their lunch stop, by a native village, Johnny removed a board from a fence to build a cooking fire. As he heated water for tea, one of the "moral birds" alit on the fence and began to make his call. "Do not steal! Do not steal!"

"Johnny," said George. "There's that bird again."

Johnny looked up and grinned.

"He's telling you not to steal wood for your fire."

Johnny looked momentarily stricken, then laughed.

"Do not steal! Do not steal!" George did his best imitation of the call.

Johnny tried to laugh again, but his face betrayed his concern.

"It's just a cultus chicken," he said, finally. "Just a fool hen."

∽

In late May, 1877, George was readying to leave Kamloops, B.C. and ride south with five pack and four riding animals, and three men. His crew for the season consisted of Jacinto, the packer, a Lytton native called Johnny, who would do the cooking, and a white man named Douglas MacFarlane, who would be his general assistant.

As no good map of the area existed, it would be necessary to keep a careful running survey during the entire summer, in which bearings would have to be taken from point to point, and distances estimated by the time occupied in travel. He had been unable to find a suitable scientific assistant, and so the whole of this work devolved upon George, in addition to his geological, botanical, photographic, and ethnological work.

The train of horses climbed the grassy slopes of the Thompson River Valley and entered the high country of the Thompson Plateau. This country was much more open than the country of the past two seasons, with rolling grasslands broken by open stands of ponderosa pine and aspen: much easier country to survey than the windfall-choked forests of the central Interior.

This was also a landscape that had been shaped by ice. A layer of moraine had been laid down over the bedrock formations, so that along riverbanks and escarpments George could easily read the layers associated with glacial deposition, volcanic action, ancient river gravels, and granite bedrock.

∞

With the aid of a local native guide, George explored coal and iron formations in the mountains along the Coldwater River, and, on Iron Mountain, came upon erratic boulders and bedrock striations at an altitude of 1585 metres, evidence for an ice sheet that would have been nearly 1100 metres thick.

As they ascended the valley of the Coldwater River, ponderosa pine gave way to lodgepole pine and Douglas fir, and George walked among blooming fireweed and chocolate lilies as he paced out his surveys. They left the Coldwater and entered the drainage of the Coquihalla River, where the landscape steepened, the valley narrowed, and the bush grew thicker. They were close enough to the coast, now, to see the effects of the coastal rainfall.

From the quiet town of Hope, George and his party turned east and rode towards Allison Pass and the Similkameen River. Below the cliffs of the Sumallo River canyon, they rode through a grove of rhododendrons in bloom, with magnificent pink flowers nodding thoughtfully in the afternoon heat. Up the side canyons George spotted snow-clad mountains poking their summits above the treeline. At Princeton, he found prickly-pear cactus in bloom, a sure sign he was back in the dry Interior.

The long vistas of the Interior were both inspiring and disheartening. At a rest stop on a bluff overlooking Okanagan Lake, Douglas said to George: "It's a beautiful sheet of water, isn't it?"

"Aye, indeed it is," George replied. But it wasn't the water that impressed him most. No, it was the white silt bluffs along the sides of the lake, which stretched northward from the south end for eight kilometres or more. Such massive silt deposits could only be the settling out of some great glacial lake, and the terraces along the tops of the bluffs suggested ancient beaches.

George sighed. "The ease with which great distances are seen in this country, and the long time it takes to overtake them is sometimes discouraging."

The weather grew hot and the stones became too hot to handle with bare hands. Evenings brought charged, muggy air as thunderheads billowed atop the hills and lightning crackled until after dark. Spot fires could be seen as plumes of smoke on the blue hills during the day.

An afternoon bath in the lake provided a welcome and relaxing break, and as George splashed about in the shallows he was visited by a garter snake, about three feet long. George stared back at the tiny-eyed snake, brown with dark spots down its back, before it swam away as graceful as a reed.

∽

As they approached the settlement of Westwold the horses began to blow and toss their heads, balking at something unseen.

"Maybe a bear nearby," said George. "Keep your eyes open." He drew his rifle from the scabbard and laid it across his legs.

Jacinto's horse reared and he cursed it in Spanish. All eyes in the party scanned the forest for movement. There were still a few grizzlies in the country.

As the buildings at Westwold came into sight the horses' agitation increased, and moments later George saw the reason for it. In the main corral stood three camels with their lower jaws working lazily as they stared at the approaching party.

A man stepped out of the barn and came to greet them. "I'm Henry Ingram," he said, catching the bridle of George's horse and stroking its muzzle. "The camels tend to spook 'em a bit."

"So it seems," said George. "I'm George Dawson, with the Geological Survey."

"Welcome."

"Are the camels yours?" enquired George as he dismounted.

"Yep," said Ingram. "Only three left, now. All females, unfortunately."

Ingram invited the men to spend the night, and George accepted. Jacinto tethered the animals at a little distance from the corral while George and the others set up camp.

After dinner, Ingram joined the men by the fire, and George could not resist asking about the camels.

"I bought them during the Cariboo gold rush," Ingram explained. "Eighty camels from Siberia. They were sent on board two ships to San Francisco. One of the ships, though, got stuck in the ice in the Bering Sea and the animals all starved to death. Twenty-four made it to San Francisco. Nearly lost my shirt.

"Then, when I put them into service as pack animals on the wagon road, we started having the kind of troubles you fellahs had today. Horses and mules were frightened by the camels and refused to pass them on the road. That caused some problems, I can tell you, and made me a little unpopular. Which was a shame because camels are excellent pack animals."

"But they couldn't work alongside horses and mules," said George.

"That's right," sighed Ingram. "Finally, the camels were prohibited on the wagon road due to the problems they caused among the conventional teams. So I moved them here. A few got free and ran off, grizzly got another, several died of old age, I suppose. And now all I have is these three females. It's a shame I don't have a male to breed them with."

"Why would you want more of them?" asked George. "Don't they just cause problems with other stock?"

"Yeah, but I'd say it's just a matter of them getting used to each other. The camels are real good pack animals, they withstand the winter just fine. They're tougher than cattle, and their hair is worth a dollar a pound in New York."

∞

Back in Kamloops, George began preparations for an exploration of Shuswap Lake by canoe. Finding a canoe, it turned out, was not easy. Mr. Chase had a large, old dugout that George was welcome to use, but it needed much repair. Smaller, two-person canoes, which George preferred, were scarce.

He visited the native village with Douglas and Johnny to see if he might find canoes to rent or purchase, but the prices being asked were too high. "Looks like nobody wants to be without a canoe," said Johnny.

They returned to Mr. Chase's for a second look at the cracked hull of his old dugout. "Well," said George, tipping his hat back to scratch his head, "She'll need some work."

With borrowed tools, the men scraped away the old patches and rotted wood, then gathered pitch, strips of tin, and a blanket for the patches. While Johnny smeared pitch into the cracks, Douglas followed behind and laid strips of blanket into the pitch. Then George nailed strips of tin over the works. They strengthened the thwarts with fresh wood and shaped new paddles out of planks. By afternoon they were ready to launch the canoe.

"We'll call her the *Pseudomorph*," George announced. "Just like the mineral that takes the appearance of another that it has replaced through chemical action, so too has this canoe taken the form of another canoe through various chemical actions applied by ourselves."

"A fitting name," Douglas agreed.

Once in the water, though, the *Pseudomorph* began to take on water. "She'll swell up and close the cracks," said Douglas.

"No," said George. "She needs more patching."

They hauled the *Pseudomorph* back up the beach, built a fire, and leaned the canoe up on edge to dry the hull. They repeated the entire process with pitch, strips of blanket, and tin, and by dark the *Pseudomorph* was floating lightly on the water, ready for her first voyage in many years.

The next day George hired another native paddler for the trip, named, appropriately, Noah. After a quick stop to gum some more cracks, they were underway.

∽

The shores of the lake were dotted with the camps of many native families. "They are waiting for the salmon to come," explained Noah. "That's why you couldn't find a canoe in the village."

A heavy thunderstorm in the afternoon chased the *Pseudomorph* off the lake and onto a gravel beach. Quickly, they turned the canoe over and stowed all the gear under it, then took refuge from the rain under the cedars and pines along the shore. While the others sat watching the rain fall, George poked around in the bush behind the beach and discovered ripe red huckleberries and black raspberries. A rocky point jutted out from one end of the beach, and he scrambled onto the wet rock to examine it. On a vertical face just above water level he found horizontal grooves in the stone where a river of ice had ground its way past.

The next afternoon, George was bucking out a specimen when splinters of schist flew into his eye. "Aargh!" He dropped his hammer and cupped a hand over the eye. After several moments of waiting for the pain to subside, he removed his hand and closed his good eye. His vision was so blurred he could not make out his hammer at his feet. Sharp pains pierced his head each time he blinked.

"Are you okay?" asked Douglas.

"I think so."

George got out his mirror and tweezers and, one by one, gingerly extracted the tiny splinters.

Douglas was concerned. "Can you see out of it?"

"Not very well, right now. But I've got all the splinters out. It'll be better in a while."

Sure enough, after a couple of hours of rest, George was able to join the men fishing off the point, and he even landed a couple of trout for dinner.

∞

As they returned through Sicamous Narrows, they met two canoes coming quickly up the lake.

"Hi-you salmon! Hi-you salmon!" the paddlers shouted as they approached. Many salmon! Many salmon! They braked hard with their paddles and caught the gunwales of the *Pseudomorph*. "Chief Louis fell off his horse and nearly killed! Hi-you salmon at the little lake!"

Their news delivered, the messengers pulled away, their paddles churning up the water. The *Pseudomorph* seemed to glide along a little easier and a little quicker after that, as her crew anticipated the sight of the returning salmon.

At the end of the lake George steered the *Pseudomorph* into the slow water of the river between the lakes and let her drift. A steady stream of salmon passed over the shallows, under and around the canoe quicker than could be counted. For an hour, George and his crew drifted over the stream of silver bodies. Jumpers splashed and rippled the surface with their fins as they returned to their rivers of birth.

"Incredible," breathed Douglas. "How many would you estimate?"

George leaned an elbow on the gunwale. "I'd have to say over a thousand in the time we've been here. Possibly more. I've never seen anything like it."

"Nor I," said Douglas.

"What's amazing is that so many make it this far, after running the gauntlet of nets and canneries at the coast, and then swimming hundreds of miles of rough water up the Fraser."

Johnny and Noah prepared pitch sticks and spears that afternoon, and that night the lake was dotted with the lights of torches as the native people speared fish through the night. For days to come, the shores of the lake were a hive of activity as the fish were cleaned and hung on racks to dry in the sun.

When the *Pseudomorph* nudged the shore at Chase, after two weeks and 320 kilometres, George Dawson had completed the first topographical and geological surveys of the Shuswap Lakes.

∞

They rode down the west side of Okanagan Lake as far as Peachland before leaving the valley and heading southwest towards Princeton. As they ascended from the Okanagan valley, they crossed two creeks full of bright red, spawning kokanee. The country was pleasant riding and easy to survey, with its open stands of ponderosa and lodgepole pine, and pea vine and bunch grass provided good feed for the horses.

George stopped on a rocky hillside in the sun to take a new bearing. As he opened the compass, he sensed movement among the stones at his feet. A large snake, easily over a metre long, moved slowly away from his boots. The reptile's head, though threatening-looking, was turned away. Still, George remained stock-still as the snake glided away over the rocks. It resembled a rattlesnake in every way – the head, the pattern on its back – except that it had no rattle.

Later in the day, he asked Johnny about the snake.

"Bull snake, maybe," replied Johnny. "Or he lost his rattle. It happens sometimes."

∞

By the middle of September, George was camped along the Similkameen River near Princeton. On information from local natives, George, Douglas, and Johnny rode several kilometres up Hayes Creek to inspect a fossil site.

The site was obvious, when they arrived, but on the opposite side of the creek. Crossing with the horses was impossible due to the choke of brush and boulders along the creek's edge.

"I must get across," said George, to anyone who cared to listen.

While Douglas searched for a place to cross the horses, Johnny began to remove his boots and britches. Selecting a stick for balance, he presented his back to George.

"Climb on."

George did so, gripping tightly to Johnny's back as he waded into the icy creek. Step by step he made his way across, dipping only George's boots when the water rose high up his legs.

"Thank you, Johnny," said George, as he dismounted onto a large rock. Johnny nodded.

George wasted no time examining the rock face. It was a good exposure, with plant and insect remains showing on the surface. He chipped away with his hammer as rain began to fall, quickly soaking through his linen coat and straw hat. Johnny took shelter by crouching under a fallen log nearby, still grasping his stick.

George barely noticed the rain as he chipped loose fossils of water skippers and fish scales – the first vertebrate fossils he'd found in the area. It was a significant find. Johnny soon began giving gentle hints that George was taking too long and that he was getting cold. "Wake sia kopet sun," he said. The sun is nearly finished.

Once satisfied he had enough samples, George wrapped them in his handkerchief. "Okay, Johnny," he said. "We've done enough work here. Let's go."

Again, he climbed onto Johnny's back and crossed the creek to where Douglas was waiting with the horses. They rode back to camp in the dusk and rain, arriving cold, wet, tired, and hungry.

Sifting through the fossils in his tent that night, George was struck by how similar the fossilized water skipper was to the present-day form. It well might have skipped across the water yesterday instead of countless years ago.

∽

Descending again to the Fraser River, they headed up the canyon road to Lytton. On the north side of the confluence, George and Douglas visited an old burial site that appeared to have been long neglected and allowed to go to ruin. Skeletons lay exposed, surrounded by scattered bones and implements, a few of which George placed in his coat pockets before leaving.

As he sat by the fire after dinner, his mind kept returning to the burial site across the river. There were many good samples there that were well suited to the museum exhibits he and his father were assembling at the Redpath Museum at McGill College.

He turned to Douglas, who was lounging against a saddle. "I'm going back over there," said George.

"Where?" asked Douglas.

"To the burial site."

Douglas pulled himself upright. "What for?"

"More samples."

"Do you think we should?"

"I don't see why not," said George. "The site is all but abandoned, and nobody seemed to mind us being there today." He got to his feet. "Are you coming?"

"Sure," said Douglas.

George dug into a canvas duffel and pulled out a gunnysack, and the two walked quietly across the bridge and up the bank to the site. A strong, cold wind blew down the valley, hissing through the pines and whipping the men's pants against their legs. A quarter moon provided a wavering light through drifting shreds of cloud.

"Just the skulls," said George, as he bent to retrieve a hollow-eyed shell.

In no time they had located seven skulls and placed them gently into the burlap sack. "Okay," said George. "That's all I want."

Next morning, George and Douglas hammered together a wooden crate and packed the skulls into it, padded them carefully with straw and rags, and shipped the crate off to Victoria on the first passing stagecoach.

"The village seemed deserted except for the sound of drums."
Skidegate Indian Village, 26 July 1878.
Exposure: stop no. 4. 13 seconds.

7

1878: To The Charlottes

> The mountains & hills everywhere rise
> steeply from the shore & there appears to
> be no arable land. Scarcely indeed any soil
> properly so called anywhere. The trees –
> among which there appears much dead
> wood – grasp the almost naked rocks.

George and Rankine turned the corner onto
Government Street and hoofed it along the water-
front as they headed back to their rooms. "It's not like
San Francisco," said Rankine, George's younger brother.

"How do you mean?" George asked. He had convinced Rankine to join him as his assistant for the

summer, before entering his studies in medicine at McGill Medical School. They'd been tramping about the harbour of Victoria, B.C. all afternoon, trying to find Captain Douglas, whom George had engaged for the summer's voyage to the Queen Charlotte Islands.

"Well," said Rankine. "The number of good-looking girls in this city in no way compares with their abundance in San Francisco."

"Hmm," George replied. "I suppose that may be so."

"You don't even notice, do you?" said Rankine, disappointed in his older brother.

"Notice what?"

"Girls. The lack of them."

Actually, George hadn't noticed, so preoccupied was he with locating Captain Douglas and making final preparations for departure for the North Coast. Douglas was a day late. *One more day*, thought George, *and then I'll have to find another ship.*

"And Father is always so proud of your powers of observation. He'd be disappointed, Georgie, old boy."

George allowed himself a smile. "You may be right, Rank. But where we're going this summer you'll see next to none, so enjoy the pretty faces while you can."

He took Rankine's elbow and steered him toward a gangway. "Down here," he said. "I want to have a look at this schooner, the *Wanderer*."

At twenty-five, John Sabiston was young for a captain, but his father had been a well-known pilot on the coast and John had learned his trade at an early age. "She was built as a pilot boat," he explained to George and Rankine. "Plenty o' beam. Strong as a horse."

George inspected the bare masts and rigging. "Steam power?"

"None needed," said Sabiston proudly. "She sails handsomely with barely a breeze."

"Where are your sails?" Rankine asked.

"A new set's being sewn up as we speak," replied Sabiston.

"How soon can she be ready to sail?" asked George.

"Inside a week, tops," Sabiston assured him.

When Captain Douglas had still not arrived the following day, George hired Captain Sabiston and the *Wanderer*. Two weeks later, the sail making delayed by the Queen's Birthday celebrations, the *Wanderer* was finally ready, and on the 27th of May, George and Rankine and the crew sailed out of Victoria, heading north via the Strait of Georgia.

Twelve days after leaving Victoria, with much time lost to-ing and fro-ing about with the tide and intermittent winds, the *Wanderer* dropped anchor at Bella Bella. A small Hudson's Bay post meant a rare opportunity to send and receive mail.

That evening, Rankine lounged and smoked on the aft deck while George jigged his fishing line, more out of curiosity than a need for food. George wasn't paying too close attention to the task as he remembered his first encounter with a canoe full of Haida Indians that afternoon. Three days out of Skidegate, they were on their way to Victoria, eight men in a dugout canoe of about 7.5 metres. The Haida were strong and confident young men on their way to the city with a cargo of dried fish and fish oil for trade. George was struck by their fair skin and fine features; they did not look like the fierce warriors the stories described. These men possessed obvious physical strength, but they were friendly and good-natured as they conversed readily in broken English and Chinook.

George suddenly remembered the fishing pole and began to haul in. There was a weight on it, as though he'd snagged a clump of weeds. He leaned over the rail to see a huge sunflower starfish rise to the surface on the end of the line. As George tested its weight, preparing to haul it up to the deck, the starfish slid off and sank slowly into the depths. George pulled in the line to find the head of a silver ratfish, a parrot-beaked little fellow who'd taken the hook and then been bitten off by some larger fish, probably a small shark. The starfish had promptly begun to feed on what was left until George hauled it up.

"Did you see that, Rank?"

Rankine pushed himself off the aft rail. "See what?"

"The starfish eating the ratfish that had been eaten by another fish. The food chain in action," said George.

"Missed it, old boy."

In the dimming glow of sunset, George decided to stow his fishing gear and call it a day.

"Tomorrow we cross to the Charlottes," he said to no one in particular.

∞

Two days later, having sloshed about on the Pacific like a cork in a bathtub, they could see the spinal humps of the Kerouard Islands rising out of the ocean like the backbone of some giant creature. A gale blew up as the *Wanderer* and crew closed on Cape St. James, and for another night they tacked offshore, waiting until daylight to find the entrance to Houston Stewart Channel.

From the anchorage at Rose Harbour, George, Rankine, and two crewmen set out in the rowboat to begin exploration of the surrounding shores. George's aim in the Queen Charlotte Islands was to conduct a general exploration and mapping of the islands – current Admiralty charts gave only broad outlines, with no detail or naming of most of the bays and channels – with particular attention to potential resource exploitation. There were also some failed coal-workings, near Queen Charlotte City, that he planned to examine.

The south end was steep country, where the hillsides ran up abruptly from the sea, leaving little workable land. The sea air seemed to magnify the mountains and trees in a way that was dreamlike, as if the place were not quite real, or somehow more than real, if such was possible.

They rowed westward as far as Cape Fanny, encountering scores of Pigeon Guillemots and Tufted Puffins, before turning and setting the sail for the run back to Rose Harbour. As their slender craft leaned with the breeze and clipped along, Rankine pointed back towards Anthony Island.

"Look! Smoke!" he shouted.

"Probably a village," George shouted back.

"D'you think they're signalling us?"

"Possibly," said George, and the others nodded as the breeze freshened and they leaned into the gunwales to trim the boat.

Shortly after supper a canoe full of young Haida men came alongside the schooner. Several of them had just returned from Victoria and were proud of their earnings and new western clothes.

"We're having dancing at the ranch," said one. "C'mon over."

"Plenty halibut. Seal skins for trade."

George settled himself against the rail. "No, thanks. We're not in the trading business. Do you know of any unusual rocks in the area?" he asked.

"Rocks all over the place," came the reply, accompanied by chuckles.

"You comin' over to see the dancers?"

"No, but thanks for the invitation," replied George.

Their curiosity satisfied, the men eased the canoe away from the side of the schooner and with a grunt and a "Hoo!" began to pull water with their paddles. In minutes they had rounded the point and were gone from sight.

Two days later the *Wanderer* hauled anchor, and George set out in the rowboat with Rankine and Williams, one of the GSC crewman. Sabiston and crew sailed northward to the next bay, which George would later name Carpenter Bay, after the English naturalist, William Benjamin Carpenter. The rowboat travelled close to shore, making a running survey as they went, with George taking bearings and measuring distances between points of land.

Just north of Carpenter Bay, George paid tribute to his brother's work as his assistant by naming a pair of small, wooded islands "the Rankine Islands." It fell to George, the first white man to map the area in detail, to provide place names, and he did so by paying tribute to many of his teachers, mentors, and family: Lyell Island, Huxley Island, Ramsay and Murchison Islands, Dolomite Narrows, Selwyn Inlet, to name a few.

The weather grew wet and windy for the next week, soaking the men by day and causing the schooner to drag anchor one night in Skincuttle Inlet. George categorized the rain, in his journal, in three types: heavy, heavier, and heaviest. The low cloud and mist often made it extremely difficult to sight from one point to another and estimate distance, which hampered the survey work. At night, aboard the schooner, one of George's regular tasks became the drying of his socks and notebook.

∞

North of Lyell Island, George and his crew came upon the village of Tanu. Four large dugout canoes were pulled up the beach at different heights, awaiting their next voyage. High up at the top of the beach sat several smaller dugouts, used for food gathering and short journeys. Fronting the beach was a row of twelve large houses typical of the style built along the coast. In front of each house stood a carved pole, each with a different set of figures displayed up and down the length of the pole. The doors of several houses were through an opening in the base of the pole.

As they eased the bow of the boat onto the pebble beach several natives came down to greet them. "Is Chief Klue in the village?" George asked, using a mix of English and Chinook.

No one in the greeting party responded, as if considering the correct answer, until a young man said: "Come."

George and Rankine followed along the front of the village to a house with a grand carved pole. Inside, they descended a set of steps to find themselves in a large rectangular depression, all covered with cedar planks except for a square area in the centre, where the fire was set. On the surrounding tiers were arranged the bedding, possessions and food boxes of the chief's family.

In the dim light and smoke, George could make out several men dressed in ceremonial blankets and sitting on the far side of the house. In the darkness of the higher levels, several women squatted on mats. One of the men, wrapped in a finer blanket than his companions, was obviously the chief. Two boxes were brought out and placed near the chief's side.

"Who is tyee?" the attending man asked. Who is your chief?

"I am," said George, and was directed to sit on the box nearest the chief.

"Welcome to Tanu," said Chief Klue. "I am Chief Klue."

"Thank you, Chief Klue," said George. "I apologize for arriving unannounced."

Chief Klue waved a hand. "That's okay. We are expecting our Skidegate friends to arrive for dancing and feasting, but you'll do," he said with a wry smile.

George learned that a new house was to be raised the next day, hence the visitors from Skidegate. He promised to come back in a few days for a proper visit, then he and Rankine returned to the boat and began the long pull back to the schooner.

Two days later George and Rankine returned to Tanu for a longer visit. Seeking information about possible coal deposits in the area, George was told of a chain of islands near the hotsprings where a thick, black substance oozed from the pebbles on the beach. He later named them Tar Islands. Before leaving Tanu, George presented Chief Klue with a gift of a pound of tobacco.

"One more thing, if I may," said George as they walked back to the boat with the chief. "Would it be okay if I took a photograph of your village?"

"A photograph?"

"Yes, a picture. I have an instrument that –"

"Oh, I know what that is. You white men and your machines. Sure, you can make a picture of my village," he said with a generous wave.

The last week of July found George and the rowboat crew camped on a sandy beach east of Alliford Bay. At Skidegate village, across the inlet, there were festivities in progress and, as they ran their survey along the south shore, George and Rankine had remarked on the number of canoes littering the village beach.

They built a large fire on the beach for warmth and as a signal to the schooner should Sabiston have decided to round the spit in the evening. After dinner, satisfied that the schooner lay at anchor somewhere safe, the three men decided to row across to the village to see the Haida dances.

When they landed in the dark at the south end of the beach, the village seemed deserted except for the persistent sound of the drums somewhere within it. Scrambling over rocks and the canoes lining the beach, the three men followed the sound of the drums to the house where the dancing was taking place.

They gathered by a blade of light that escaped the door of the house. Inside was the hum and throb of the dancers, drummers, and the watchers. "Shall we, then?" said George, and without waiting for an answer pushed open the door.

A glare of light flashed out, illuminating the three white men as they stepped inside and found themselves standing in the midst of men and women in ornamented capes and robes, cedar bark head-dresses, and eerie wooden masks resembling mythical birds and beasts. It was like walking into a wild dream.

George slipped through the dancers and crossed the open floor to the far side of the crackling fire, followed closely by Rankine and Williams. As they seated themselves among the onlookers an old man with a grey wisp of whiskers on his chin tapped Rankine on the shoulder and said something to him in Chinook. "He wants us to remove our hats," George translated as he set his own on his lap.

The big house was similar to Chief Klue's at Tanu, except that the floor was not excavated. Planks covered the entire floor except for the fire pit in the centre, which had a sparkling blaze built up for the occasion and was the only source of heat and light. Smoke hovered under the roof beams and escaped through wide openings in the roof planks. The family's belongings had been heaped along the walls to make room for the audience and dancers, who filled the house to capacity.

From where he was seated George had a fine view of the dancers, about twenty of them, gathered against the front wall of the house. Nearly all wore headdresses, variously constructed of cedar bark rope ornamented with feathers and shells, and in one case decorated with the bristling whiskers of a sea lion. Their cedar bark capes were ornamented with tassels, shells, feathers, and fringes, and each dancer carried a sprig of fresh spruce.

A man standing on a box in the middle of the back row began to sing and chant, keeping time with a long, thin stick. A drummer joined in with a double-time beat – tum tum, tum tum, tum tum – and the dancers began to sing and chant and shuffle in time, the pitch rising and falling according to the direction of the man on the box. Several dancers shook rattles as they bobbed and weaved in time to the drum.

The singing swelled to a crescendo, with the rattles shaking madly while dancers spun and jerked their bodies as though in battle. The house was filled with this great noise until the master of ceremonies gave a sign and all stopped with a loud "Hoo!"

George glanced at Rankine, who sat completely absorbed by the spectacle, and nudged him with his elbow. "It's not like San Francisco, is it?"

∞

At the northwestern corner of the islands, George and Rankine hiked across a narrow peninsula to the west coast. Though little more than a kilometre long, the trail led them over and under moss-covered logs and through thickets of salal and salmonberry. All the brush was, of course, soaking wet from the frequent showers, and George and Rankine arrived on the sandy beach in wet clothes.

"If we keep moving we'll stay warm," said George, striking out southward along the sweep of sand.

He was quickly mesmerized by the great, green waves that marched shoreward and arched up, fringed with little rainbows as their edges became fretted and misty, then crashed down in huge piles of white foam with a roar that was magnificent and humbling.

George contemplated this illustration of the power of the water to shape and erode the land. The sum of the work done left him without explanation. A bit of verse from Tennyson was all he was able to summon: "Break, break, break, on thy cold grey stones, O Sea! And I would that my tongue could utter the thoughts that arise in me."

Down the beach, they came upon thousands of gooseneck barnacles, *Lepas anatifera*, washed ashore by the recent storm. As George added his notes to the map that evening, he named the bay, Lepas Bay.

George and Rankine hiked to the end of the beach then carried on over the rocky shore another five kilometres to Lauder Point. Here, George took bearings on Cape Knox to the north and Frederick Island to the south, and made a sketch of the rocky coast with its cliffs and pinnacles and dramatic, crashing reefs.

They hiked back only to find the rowboat high and dry on the beach in Bruin Bay. The crew, which had been ashore all day cutting wood and filling water casks, had forgotten to move it down the beach with the falling tide. A native man with a canoe, who happened to be in the bay, paddled George and Rankine out to the schooner to change out of their wet clothing.

∽

From the Charlottes, the *Wanderer* crossed to the mainland, where Rankine and George parted ways at Metlakatla as Rankine boarded the steamer *Grappler* to speed his return to Victoria and Montreal and his studies.

On the voyage home aboard the *Wanderer*, George stopped again at Bella Bella, where he met a man named Hain-chit. He and George walked along the beach and Hain-chit shared with George the stories of his people's origin, and how they obtained fire.

When he was finished telling the stories, he fell silent. The men paused at the end of the beach to watch salmon finning in the shallows, and when they turned back towards the village, Hain-chit began to speak again.

"My people talk much about how few we are now," he said. "Long ago we were like the trees. Great numbers everywhere. We fought each other, just like the whites do, and some were killed. But always more were born and the country teemed with people the way the rivers fill with salmon every fall.

"But now, the whites have come into the country and everywhere the Indians chaco mamaloose, chaco mamaloose..."

George understood the Chinook phrase, "chaco mamaloose." He'd heard the same thing said by the Interior native people. It meant: they have died.

"Soon," Hain-chit continued, "there will be none. This village was once great. Now it is weak. There is still plenty food, plenty fish, and we have things from the white men we did not have before. Tools, guns. We do not fight among ourselves like we did before, and the whiskey is not around so much now.

"But yet we die. I don't know how to explain it," sighed Hain-chit. "Klunas saghalie tyee mamook." I don't know what God is doing.

8

1887: Rivers of the Yukon

Possible hostile aspect of Pelly Indians adds additional uncertainty, and makes the whole affair very little like a pleasure trip. If we succeed we shall do it in spite of many difficulties.

On the second morning aboard the *Alaskan*, the little steamer making its first voyage of the season up the Stikine River, George Dawson let himself relax a little. For the first time since leaving Victoria – ten days later than planned – things seemed to be going the way he liked.

The steamer was loaded to capacity with its cargo of Dawson's Geological Survey crew and gear, and

"A very toilsome journey." Looking Up Pelly River, Yukon, 4 Aug 1887.

Chinese miners bound for Cassiar. The *Alaskan* had no cabin space, which meant they tied up to the bank at night so the passengers could go ashore and pitch camp. For George, the survey expeditions began when he first pitched his tent and wrapped himself in his blankets and buffalo robe to sleep.

The spring melt had begun, swelling the river to make swift water where there was usually slow, and riffles where it was usually flat. In late afternoon, as the steamer churned its way up yet another stretch of fast water, she lost power for a moment, and before she could be got underway again, the *Alaskan* began to drift backwards with the current.

George lifted his head from his note making when he felt the momentum cease, and was alerted to danger by the shouts of the crew members. He rushed to the rail to see the scenery passing backwards as the steamer drifted idly, her bow swinging slowly towards midstream. He felt the grind of the rudders as the boat ran aground on a gravel bar. The bow continued to swing downstream, passing the stern, and for two miles the *Alaskan* cartwheeled thus downstream, in serious danger of smashing the sternwheel against the trees and rocks along the shore. Captain, crew, and passengers watched helplessly until, finally, the steamer drifted into a large eddy and a line was heaved ashore and secured. *Two steps forward, one step back*, thought George.

George and his crew went ashore and made camp in the snowy woods while the crew plunged into the icy water to unship the bent rudders.

Next morning, the men built great, roaring fires and heated the rudderstocks until they glowed red and menacing. They removed the glowing stocks from the fire and, while a gang of men strained on the end of a long lever and others pounded furiously with heavy sledgehammers, the rudderstocks were straightened one at a time. By afternoon the steamer was reassembled and they were underway.

The Captain ordered the safety valves fastened down to increase the steam pressure and power to the wheel. *A dangerous practice, that*, thought George, but the only practical way to move upstream against the powerful current.

∞

After six days of running the steamer to absolute capacity, lining up one rapid and bouncing the hull off a rock in another, the *Alaskan* arrived at Telegraph Creek, the terminus of its run.

George learned that Dease Lake, eighty kilometres away, was still covered with ice and probably would remain so for another week. He sent crewman Davey Johnson and a local man ahead to begin sawing lumber for boats. George, McEvoy, "Brick" Lewis, Jacinto, and Larry, a Tahltan assistant and cook, would follow as soon as the pack mules arrived from Glenora.

While McEvoy surveyed the surrounding hills, George spoke with the local Tahltans, trying to learn their language and gather information about the country. A confusing mixture of Chinook and English, as usual, was the language of communication.

Although the lake was still thick with ice, along the shore willow catkins bloomed and leaves began to open on willow and poplar trees. It was now the first week of June; the lake was still frozen and the boats were not ready. Again, George was visited by a feeling of the season slipping by too quickly.

∞

A party of miners arrived, having come up the Stikine by canoe, and one of them, a man named Campbell, sought out George with a note from Mr. Gallbraith of Wrangell. The note stated that a report of trouble between Indians and miners in the Yukon had been confirmed. Two whites and four natives had been killed in a scrimmage at Harper's, near the mouth of the Stewart River.

"For what it's worth," said Campbell, "the fellah that brought the news out is not a man of good character. Though most folks is crediting him with the story as he gave the names of the whites killed, including a son of Harper's."

"I see," said George. This was not good news. "Do you know what the skirmish was about?"

"Grub," said Campbell. "Supplies are scarce, and Harper refused to sell to the Indians. Hence the attack. The culprits apparently fled up the Pelly, or in that general direction. For reinforcements, some say."

"We could be running straight into a hornet's nest on the Pelly," George observed. "And if what you say is true, they'll want the lives of two more whites to balance the ledger."

"That could be," said Campbell. "If it's all true."

"A big 'if,'" said George. He was worried, too, about Bill Ogilvie, who was to travel down the Yukon

River and meet George at the confluence of the Yukon and Pelly before wintering over in the area.

By morning George had decided what he had to do. Given the reputation of the miner making the report and the chance that things might be resolved by the time they reached the Pelly, at least a month from now, George decided to continue the expedition. He would seek further news at Lower Post, on the Liard River, before committing to the next leg of the journey.

∞

On the 16th of June, 1887, Dease Lake cleared, and George's party set off down the lake in their three new boats. On the 19th they began descending the Dease River, winding through mountainous country forested with black spruce and pine. Floating over shallows and gravel bars, George leaned over the gunwale to hear pebbles under the boat singing as they were tumbled along by the current.

∞

The purpose of the season's expedition was to gather information on the vast territory of the Yukon. Little was known about its geology and resources, although miners had been trickling into the Yukon for several years, seeking gold. The map George carried had been drawn by Robert Campbell, who had traversed the area in 1840. Since then, no other mapping or surveying had been done.

Four days later they emerged from the mountains into a low, dreary-looking country under showery skies and cold rain. The Hudson's Bay Company fort on the Liard, at the mouth of the Dease, was a bleak and bare collection of log buildings. Never having seen so many whites before, the local Tlingit, or Stikine, people were very curious, and they checked over the outfit, paying particular attention to the canvas-covered, wood-framed Osgood boat and George's inflatable pillow.

With two local "Stick," as the Tlingit were called, hired to travel by foot and to hunt for the expedition, Dawson headed up the Liard River. The party now consisted of Dawson, McEvoy, Johnson, Lewis, Captain Jim and Charley, both Tsimshian boatmen hired in Wrangell, Jim's wife, and the two Stick hunters.

At the top of the Liard canyon, George took a latitude observation and discovered they were 2.4 kilometres north of the B.C.–Yukon border. In a cloud of mosquitoes, he cut a wood post, stood it in a cairn of stones on a prominent point overlooking the river, and engraved the post with the latitude: 60 deg. 1'6".

The work of poling and tracking the boats up the riverbanks was hard and tedious, and by the end of each day the men were hungry enough to eat a horse. Already George was concerned about their provisions holding out. So far, the Stick hunters had not brought in any game. In fact, they hadn't been seen for two days. Were they lost, or had they deserted? That night, George had the other native men set fire to a large spruce tree as a signal smoke for the missing Sticks. The spruce, when fired, made a tremendous roar and produced a great column of smoke that ascended a couple of hundred metres in the calm evening air.

The men set their nets at night and caught pike, whitefish, and steelhead trout to supplement their diet of biscuits and bacon. What they needed was a moose.

The next day the Stick natives rejoined the party, and two days later they informed George they weren't going any farther. George was puzzled by this. Had they heard something of the trouble on the Pelly that he hadn't? Unable to stop them, George watched them leave.

Leaving the Liard, George's party began to work its way up the Frances River. The fishing got better, as each day they caught several large trout, but the poling and tracking of the boats was hard, telling work for the men, and frequent cursing became necessary.

It was no small relief to reach the still water of Frances Lake. "My God," George breathed. "Tell me if this isn't a beautiful sheet of water. Those mountains must reach four thousand feet above us."

"Indeed," McEvoy agreed. "And fresh snow on the peaks."

It was a reminder of how short and tenuous summer could be in the North Country. They poled and rowed up the winding shoreline, past tiny coves and gravel spits, searching for signs of native camps and finding none. Jim shot a moose, and that night the men enjoyed all the moose steak they could eat. Jim's wife and Charley hung the remaining meat over the dying fire to cure, and laid leg bones in the coals to roast the marrow.

"I can't help thinking," George said to McEvoy as they settled into their blankets. "That the Sticks from Liard may have heard more about the troubles with the whites than we did."

"That's why they turned back?"

"Possibly," said George. "Possibly. It just adds more uncertainty about what lies ahead for us."

"Quite true," McEvoy replied. "You thinking of turning back?"

"Not for a moment," said George.

"Didn't think so."

⬡

After locating the mouth of the Finlayson River, George decided to call the next day Sunday, as they had been on the move every day since leaving Lower Post on the Liard.

After a day of rest, tending to personal effects, and talking over weights and arrangements for the overland trip ahead, George loaded a boat with several days' supplies and set out with Lewis and Charley. They explored the east arm of Frances Lake, both to confirm Campbell's map and to look for local native people to hire for the portage to the Pelly. The map turned out to be accurate enough, but they found no recent traces of camps. In previous expeditions George had relied on local native people as sources of labour and information about the country and the best routes of travel. It was unsettling to have to operate without them.

At the camp at the mouth of the Finlayson, they found that McEvoy, Johnson, and Jim had completed the cache for the gear and food they would leave behind. McEvoy had also done some exploration, and he explained to George that there was a canyon above, up which it was impossible to haul a boat. The Osgood boat would have to be dismantled and portaged around the canyon before being reassembled and hauled the remaining distance to Finlayson Lake.

After six days of wading through swamps, hiking through deep, spongy moss, and breaking through birch willow and Labrador tea bushes, the men came upon Finlayson Lake.

"A fine sheet of water," George declared, and all agreed.

They chopped down enough spruce and pine trees to build a raft, and sailed and rowed the two craft up the lake, surveying the shore as they went.

The lake ended in a large slough, choked with big, yellow cow-lilies. Loads were trimmed again, leaving behind enough flour and bacon for the natives' return trip, and the raft was pulled apart and dragged into the bushes so the logs would be dry when they returned.

The remaining gear had been divided into six packs, each weighing, George estimated, not much under forty-five kilograms each. As the men stood eyeing the packs on the ground, George sensed a reluctance to shoulder the loads.

"Dr. George," said Jim, the Tsimshian headman. "We have never been so far from home. It gives us a sad feeling." Jim's wife and Charley both nodded in agreement.

George surveyed his tired and beaten-looking crew, their clothes torn from bushwhacking, their hands and faces lined with scratches. This trip was certainly more physically demanding than any other George had done. Johnson, in particular, was nearing exhaustion. The Tsimshians weren't about to walk away from two months of high wages, but Jim was trying to express what everyone was no doubt feeling. The men were wearing out.

"I know," said George. "We're all being tested to our limits. But we have to push on. It makes no sense for any or all of us to turn back now, don't you see?"

The Tsimshians stared at the ground, saying nothing.

"You and Charley have been critical to the success of this expedition. Without you we wouldn't have made it this far. And we need you to get through to the Pelly."

"We're close to the Pelly now," said McEvoy, waving a hand at the cloud of mosquitoes around his head. "Probably not more than a couple of days before you'll be heading home."

George breathed a silent sigh of relief and gratitude at McEvoy's support. "Quite so," said George. "And to show our appreciation for your hard work, we'll give you each a ten-dollar bonus when we reach the Pelly."

Jim looked at George, then at McEvoy. "Okay," he said, nodding to Charley. "Let's go to the Pelly."

Again, there was no trail to follow, so George took a compass bearing and set off through the bush towards, he hoped, the headwaters of the Pelly River. Two hot days of bushwhacking through gullies, ravines, and steep creek banks brought them to a watershed that flowed west. If it wasn't the Pelly itself, it was at least a tributary. George allowed himself a small moment of triumph.

They camped on a wooded bench by the river, where there was good timber for building another boat.

"I want to go back with the Indians."

George looked up from his notebook. It was Johnson. His eyes were sunken and lifeless and looked at George pleadingly. He had been tested more severely than any of them.

"Have a seat, Davey," said George. "You look done in."

"I am," he said, lowering himself to the ground beside George. "I can't go on."

"The worst of it's behind us now," said George, closing his notebook.

"I don't think I can do it." He sat forward, supporting his head with his hands.

"You'll get some rest on the Pelly," said George. "You know that, don't you?"

"Yeah."

"I really think you should push on with us. Take it easy for a bit, and you'll feel better in a few days. Think how you'd feel if you gave up and turned back.

"We all have our bad days, Davey, but we have to just grit our teeth and ride them out. It always gets better. You have to believe that."

Davey met George's eyes briefly. "I guess you know something about that," he said.

∞

The descent of the Pelly was made in the newly built canvas-covered, gum-and-bacon-fat-sealed canoe. Needing a boat that was light enough to portage through Hoole's Canyon, and one that wouldn't take long to construct, George had packed a piece of canvas the entire journey for this very purpose. Now that they were on the Pelly the concerns about the Indian troubles were on their minds again, and George wanted to be able to move as quickly as possible.

They met no native people on the Hoole's Canyon portage, but below the canyon they began to see rafts pulled up on the south bank of the river where native people had crossed the river. George kept his rifle by his leg as they travelled, and a net of tension descended over the expedition, as they knew sooner or later they must encounter Indians.

A week down the Pelly they met their first native people, the first humans they'd seen in six weeks. "Hello!" George hailed them.

The man raised his hand in silent greeting as they eased their canvas canoe into the eddy beside the dugout. The man and his son wore silver rings through their noses, cloth shirts patterned with beadwork, and canvas pants. They were neither hostile nor afraid.

George offered the man tobacco and matches and began to speak to him in Chinook, of which the man understood little. He seemed surprised to see white men this far up the river, and through a series of gestures and signs indicated that most of the people had gone up the creeks and rivers to fish, hence the rafts along the banks.

Two days later they met a couple of miners, the first white men they'd seen since Lower Post.

"Where the devil have you boys been?" asked the smaller of the two men, catching the line Johnson tossed.

"Came over from Lower Post on the Liard," said George.

"Good Lord," said the second miner. "Only one other white man's ever made that trip."

"Yes," said George. "Robert Campbell, in 1840."

"You boys as crazy as he was?"

"I doubt it," said McEvoy. "We're with the Geological Survey."

"What about the troubles at Stewart Crossing?" asked Johnson.

"Troubles? No troubles."

"We heard there was some shooting. Two whites and four Indians killed at Harper's place."

"Oh, that 'trouble.'" The miner scoffed. "That story was started by a fellah who got run out of the country for misconduct. Ain't no truth to it at all."

"You don't know how happy I am to hear that," said Johnson.

George allowed himself a little smile. Davey had bounced back just fine.

∽

At the confluence of the Pelly and Yukon rivers, George and McEvoy searched for the blazed trees and note from Ogilvie, who was to meet George's party and provide a resupply. But they found no blazed trees and no note. This meant they would have to descend the Yukon 350 kilometres to Fortymile, then return upriver and continue to the Chilkoot Pass and out.

Some passing miners said they'd seen a man who fit Ogilvie's description building a boat at Miles Canyon twenty days earlier. George also learned that Harper's trading post had been moved downriver to the Fortymile, where supplies were now insufficient. Some of the 250 miners who'd planned to winter over were having to go out because of it. This was not good news for George and his crew, but there was little they could do except carry on. He also learned that some very rich ground had been worked this summer on the Fortymile.

George selected a camp with good timber to build another boat for the ascent. Next morning the men chopped down several trees before finding one that was sound, and by late afternoon they had whipsawed two planks from it.

While the men worked at the boat building, George checked over his barometers and updated his notebook. A freight canoe rounded the bend, and George paused to watch. A man stood in the stern, ruddering with his paddle. There was something about the set of his hat, the roundness of his shoulders that seemed familiar. Ogilvie!

George reached for his rifle, and fired a shot into the air. A moment later Ogilvie raised his own rifle and fired a reply. George scrambled down the bank to greet him.

"Well, well," said George. "If it isn't Mr. Ogilvie. How are you, Bill?"

"Hello, George. Very well, thanks. A devil of a time portaging all this gear. Bit of a delay."

"Ourselves as well," said George. "We arrived here yesterday."

"Really? I'd have thought you'd be much further along."

"So had I," said George. "So had I."

It was a great relief to meet Ogilvie, and to arrive here only two days apart after nearly three months of wilderness travel was remarkable.

The following day, August 14, was declared Sunday in both camps. After spending the morning tending to his personal gear, George spent the afternoon with Ogilvie, discussing the division of provisions, their maps, and the reports of good gold diggings on the Fortymile River.

Four days later George and his crew set off up the Yukon River with their new boat, a 7.5-metre double ender, built from a single tree. Along the riverbanks and surrounding hills, aspens and cottonwood trees were beginning to turn yellow. Dead and dying salmon littered the shores, and bear sign was abundant. The nights grew cooler, with frost clinging to the tent walls in the mornings. Each morning a quilt of mist lay over the river until the sun burned it off.

Almost daily, they met miners heading out for the winter, and a few heading in to tough it out and get an early start in spring. Lake Laberge greeted them with a headwind, and they made a slow survey of its eastern shore to match Ogilvie's survey of the western shore. Swamp grass along the shores turned to russet and yellow overnight, it seemed.

At White Horse Rapids, so named for the flowing white manes of the breaking waves, the men portaged the boat and gear nearly a kilometre, reloaded, and tracked it up another 1.5 kilometres to a set of skids laid down for a second, shorter portage. They made another portage around Miles Canyon and above that saw many dead salmon on the banks, half-eaten and rotting, and bear tracks up and down the banks. The stink of rotting fish and presence of bears made camping unpleasant and gave the men reason to travel quickly.

By the time they camped on the shores of Bennett Lake, the aspens were flaming yellow among the dark pines and spruce, and here and there red and orange patches of undergrowth blossomed sharply. Snow was possible any day now from the low, grey clouds that moved sluggishly across the sky.

At Lindeman Lake, George hired four packers, and on the 19th of September 1887, they crossed the Chilkoot Pass, arriving at Taiya Inlet on the evening of the 20th.

The summer's journey had covered, according to George's calculations, 2115 kilometres, and circumscribed an area of about 101,120 square kilometres. And though gold was already being mined on many rivers, it would be another eleven years before the big strikes of Bonanza Creek and the Rush of '98, and the naming of Dawson City in George's honour.

"They began to talk rapidly and brandish their shiny knives."
Natives skinning fur seal carcasses. St. Paul Island, Bering Sea.

9

1891: A Voyage to the Bering Sea

> The firing at a pursuit of the sea lions in the
> steam launch, dipping about in a rather bold
> swell which was breaking on Gull Rock and
> adjacent rocks was rather good sport, but
> the day's work as a whole only play and
> nothing to do with the seal enquiry.

Heading northwest from Triangle Island, the
Danube slowed as it came alongside an object
adrift and partially submerged. George leaned over the
rail as far as he dared for a closer look.

"It's a boat of some sort," said a crewman.

"An Indian canoe," said James Macoun.

George squinted down at the odd-looking boat. It was definitely a native boat but it was not a dugout, the usual style of boat used by Indians on this part of the coast.

"It appears to be made of driftwood and large skins sewn together," said George. "I've not seen one like it before, but it looks in every respect like the style used where we're going."

"Is it an Aleutian canoe?" asked Sir George Baden-Powell, one of George Dawson's travelling companions.

"I believe it is," George replied. "A *baidarka*. Strange, isn't it, that we should encounter one here? It has drifted a very long way."

"I wonder what sort of tragedy befell the natives who paddled it?" Sir George pondered aloud.

The question was unanswerable, but the awash *baidarka* was sufficient reminder of the harsh and unforgiving way of life on the northern oceans.

As the sun sank into the ocean, George recorded their position at Latitude 52 deg. 2', Longitude 140 deg. 29', 480 kilometres directly west of the southern tip of the Queen Charlotte Islands. They were steaming directly across the Gulf of Alaska to the Aleutian Islands and the Bering Sea, birthplace of the winds.

∞

Five years earlier, in 1886, the Americans had seized three British sealing schooners in the Bering Sea. Operating under the belief that the Bering Sea belonged to them, as they had bought all rights from the Russians when they purchased Alaska, they continued to seize foreign sealers at every opportunity over the next few years.

In the spring of 1891, the American government had given notice that they were sending several men-of-war to the North to seize any kind of vessel whatsoever that entered the Bering Sea. The British government responded immediately by forming the Bering Sea Commission and sending the members of the commission to the area to represent Great Britain's interests in the political troubles surrounding the fur seal fishery. George Mercer Dawson, now Assistant Director of the Geological Survey, and Sir George Baden-Powell, a British statesman and yachtsman, were requested to fulfill the Commission. A voyage to the Bering Sea was an easy assignment for George to accept. He'd never been to the northern oceans, and the promise of adventure was high.

The mandate he and Sir George were charged with was to undertake an enquiry into the conditions of fur seal life and the fishery, and to determine the precautions necessary for preventing the extermination of the fur seal of the Bering Sea and North Pacific Ocean.

As soon as Sir George Baden-Powell and his manservant arrived from England they were met by George Dawson and the GSC field naturalist, John Macoun, who'd accompanied George on several expeditions, and the four men stepped onto a train for the West Coast of Canada. At Vancouver they boarded the chartered steamer *Danube* and began steaming towards the Aleutian Islands.

∞

Late on the 24th of July they spotted land a long way off, a land of fog and volcanic islands shrouded with clouds. Through rifts in the clouds George caught glimpses of the tops of these high islands. Twice, he sighted plumes of smoke over the snow-patched cones that poked through clouds that flowed on urgent winds.

The *Danube* steamed on through Akutan Pass and into the Bering Sea as day broke. Rounding the steep slopes of Cape Kalekhta, on Unalaska Island, they found haven at Iliuliuk Harbour. As the crew set the anchor, George took in the bold green mountains surrounding the harbour, and the distant snowy ranges seen in bits through the roiling clouds. The hillsides were treeless, but the shrubs and grasses that grew were profuse and charged an almost unnatural shade of green. Startled by the rattle of the anchor chain, oystercatchers cried shrilly and flew in low circles about the bay. They took roost on an exposed reef, where their black bodies blended instantly with the rock, and all that remained visible were their long, red bills.

Makushin volcano and the high ranges about it lay deeply covered in snow, in some places holding small glaciers. Makushin was not a typical cone, but rounded and humpy with bold spurs and buttresses and dragon-back ridges. A plume of white steam or smoke rose steadily from behind one such buttress.

For such a remote place, Iliuliuk Harbour was a busy port. At anchor lay three British war vessels, four U.S. war vessels, one U.S. revenue cutter, and several other vessels with coal and supplies. That afternoon the *Albatross* arrived, carrying George's American counterparts, Doctor Hart Merriam and Professor T.C. Mendenhall.

∽

The North Pacific fur seal had been hunted by the native people of the Kamchatka Peninsula and Pacific Northwest for centuries. When Vitus Bering's party was stranded on one of the Commander Islands in 1742, they were the first whites to witness the astonishing invasion of the fur seals at their breeding rookeries. Soon after, the Russians began to hunt fur seals and sea otters for their valuable hides, and the commercial slaughter worked its way steadily eastward along the Aleutian Islands towards Alaska.

In 1786, Gerassim Pribilof came upon the pair of islands that now bear his name. St. Paul and St. George Islands, the Pribilofs, were, at that time, home to a million or so fur seals every summer. The supply of furs must have seemed limitless.

∽

From Unalaska Island, the *Danube* steamed north to the Pribilofs. With the sale of Alaska to the U.S., the American government had leased the islands to the Alaska Commercial Company, which was licensed to kill 100,000 male fur seals annually.

Through breaks in the fog George caught his first sight of fur seals sleeping in the water, lying on their sides with hind-flippers arched forward to meet fore-flipper in a distinctive pose.

So thick was the fog that the captain of the *Danube* missed St. George Island, and decided to carry on the sixty-four kilometres to St. Paul Island. At St. Paul the fog lifted long enough to allow them to enter the anchorage. Ashore, George and his party met Mr. Tingle of the Alaska Commercial Co., who took them for a hike to a nearby rookery.

As they neared the top of a low rise, Mr. Tingle instructed the men to lie down and crawl on their bellies. "If they see or hear us they'll stampede into the water and be gone," he said, lowering himself onto his belly to demonstrate.

From over the rise, they could hear the bawling and barking of thousands of hoarse voices, and the smell was enough to make Sir George gag momentarily. George inched forward to see the entire sweep of beach writhing with seal flesh from end to end.

"These're mostly 'holluschickies,'" said Mr. Tingle.

"What're they?" asked George.

"Juvenile males. These and a few old bulls are the last to leave. Most of the females and pups have already headed south."

"Some of those bulls must weigh five hundred pounds!" said Sir George.

"They go as high as seven or eight hundred," said Mr. Tingle. "And you'd be surprised how fast they can move."

"I'll take your word for it," replied George.

"I once saw a bull catch a man, grab him in his mouth and toss the poor beggar twenty feet."

"My Lord!" gasped Sir George.

George accepted an offer to visit another rookery, this time travelling in a *baidarka*, a skin boat much like the one he'd seen adrift at the start of the voyage. Sir George declined the offer, choosing to stay aboard the *Danube* for lunch. Mr. Tingle led George and Macoun to the *baidarka* and the waiting crew of Aleuts. The Aleuts were all dressed in their waterproof walrus intestine coats, called *kawliakas*, which, thought George, resembled badly made nightshirts. But, as they set out into the mist and spray, George's own oiled canvas jacket seemed suddenly inadequate.

As the chief steered the boat deftly along the shoreline, skirting submerged rocks with a practised eye, George breathed the marine air and tried to separate the smells of plants, birds, and mammals from the pungent cocktail of odours assailing him, one of which was the greasy smell of the Aleuts' *kawliakas*.

The paddlers halted behind a jutting point and the chief gave the signal for silence. George could hear the guttural barking and burping of the fur seals nearby, and he caught occasional sharp whiffs of the animals on the wind. Slowly, they cleared the point and George hunched his shoulders against the penetrating wind. They let the *baidarka* drift on the swell as they watched the fur seals. Thousands, George estimated, barking and humping their way across the gravel beach to and from the breaking surf and the waters of the bay, which were dotted with the heads of hundreds more.

When he climbed the ladder up the side of the *Danube* at four in the afternoon, George was thoroughly soaked and chilled to the bone. After changing into dry clothes, he found Sir George in the saloon, being served afternoon tea by his man.

"Spot o' tea, George?" quipped Sir George.

"Quite," said George, settling into his seat and raising the brim of the steaming mug to his lips. "It's a wet one. Wet enough to make me consider buying one of those *kawliakas*."

∞

Several days later, the commissioners went ashore to watch a seal-killing for native food. As they walked up the slope towards the salting shed they saw the end of the drive – a hundred or more seals being prodded along a worn track by all the inhabitants of the village. Exhausted and stricken with fear, the seals bleated and bawled as they were driven to the killing ground.

"The drive," as it was called, meant hiking across the island to the rookery beach, cutting the seals off from the water, and herding them back across the island to the killing ground. On some islands, this involved a considerable climb and descent. Some seals would die along the way, from exhaustion or being trampled, but having the seals carry their own hides to the salting shed saved the hunters much work.

George watched the seal hunters wade into the midst of the pack and begin dispatching the animals with their clubs. A good hunter could kill an animal with a single blow to the skull, a mercifully quick death after the prolonged trauma of the drive. George set up his camera and took several photos.

As the last seal was clubbed, a momentary silence descended, bringing relief from the sounds of killing. Gulls cried and wheeled in the air and descended upon the corpses. Ravens glided in and hopped among the dead seals, fighting the gulls for the seals' eyes.

At once the native hunters began talking rapidly and brandishing knives for skinning. George repositioned his camera for a better view of the skinning process, and after taking a couple of photos he joined Dr. Merriam in dissecting stomachs.

"Looks like mostly herring," George observed as he scraped the contents out onto the ground.

Dr. Merriam squatted beside him, poking through the remains with the tip of his knife. "Some squid, too," he said, impaling one and holding it up for inspection.

One by one, the seals were skinned out and their hides stacked, fur sides together and a layer of salt between the flesh sides for curing.

∞

"There!" shouted one of the crew, pointing up the sweeping green slope to the white form of a polar bear ambling along. "Pass me the rifle."

"He's out of range," said George. "Save your ammunition."

"Maybe the others'll get him."

A shot from a boat heaving in the swell would be tricky indeed. They continued to pull their way along the island's shore, past rocky outcrops and shallow bays, seeing no sign of fur seals. The crews had divided into two groups, one hiking overland and the other piloting the steam launch along the shoreline.

On a large, low reef George spotted the sleeping forms of walrus. "Look! There!" he whispered hoarsely, and immediately all rifles were raised and sighted on the massive mammals.

One of the bulls, having smelled the men, raised its tusked head, and suddenly all seven walrus were lurching towards the sea, launching their huge bodies off the rocks and into the water with surprising speed and grace.

A hail of bullets struck the rocks and water around the reef, and each time a head appeared more bullets ripped into the water around it. "Got him!" "Missed!" The men cheered and shouted with great fervour, firing now at anything moving in the water, whether it resembled a walrus or not. George sat in the heaving boat in a cloud of gun smoke, his rifle at rest between his knees. Shooting in such conditions was pointless, a waste of ammunition. He saw a patch of blood-tinged water forming among the kelp weed, a sure sign that one of the animals had been wounded. They'd never be able to track it and kill it now.

The men stopped shooting and motored along the shore again, rifles at the ready for another encounter. They saw no more walrus, but upon rounding a point sighted a second polar bear and promptly launched a hail of bullets in its direction.

"Got him!" yelled McGregor, the engineer, as the bear staggered and dropped to its knees. In a moment, the animal was up again and loping across the hillside to safety.

"Look!" Sir George stood up in the rocking boat and pointed. In the distance was the party of men on foot, directly in the path of the escaping bear.

George and the men in the boat watched as the shore party spotted the oncoming bear and raised their rifles to drop the beast with a single volley. As they skinned out the bear for its hide, the motor launch turned and began its journey back to the *Danube*, alert for the opportunity to collect a walrus.

George, though, kept his eyes peeled for signs of glaciation. So far, he had seen no evidence of scratching or deposition at any place among the volcanic islands, and he was daily more certain that the area had not been glaciated by the great Cordilleran glacier of western Canada.

∞

Clad in an official uniform with sword and attended by two Cossacks, also in military dress, Nicolai Grebnitzky greeted George and the other commissioners.

"Welcome to Bering Island, my friends," he said in a thick Russian accent. Grebnitzky was a tall, well-built man of middle age, a seal naturalist who had spent many years on the islands in the Bering Sea, working and studying alongside the seal hunters.

He led the commissioners to his office and served them tea in heavy tumblers, Russian style, then showed them the bales of skins confiscated from the schooner *Ft. Lewis*. He guided them to the new church, where an elaborate consecration ceremony was in progress, led by the Archbishop from

Petropavlovsk. George took in the odd scene inside: the ornate, Russian Orthodox stylings of the church and the native congregation dressed in skins and singing Gregorian chants.

The Aleuts here were well off. Collectively, they owned eight or nine hundred dogs, George estimated, which were all tethered on lines behind the village, and which set up a tremendous yelping and barking at the sight of the touring commissioners.

"They use the sleds all year round," Grebnitzky explained. "Not just in winter."

"What do they feed them?" asked George. "It must take quite a pile of meat to maintain that many dogs."

"Seal meat," Grebnitzky replied. "Fresh in the killing season and salted the rest of the year."

Back at Grebnitzky's quarters, George and his companions were treated to a lunch of smoked salmon, rye bread, biscuits, wine, and blueberry cobbler for dessert. While the commissioners relaxed after their meal, Grebnitzky beckoned to George to follow him into his study.

"I want you to have this, even though it is incomplete," said Grebnitzky, presenting George with a display case. In the case was a collection of beautiful samples of plants. "They are all native to these islands."

George was taken aback by the Russian's generosity. "I can't take this," he said, refusing as politely as possible. "You've invested much work in it."

"I can always get more samples," the Russian said. "You cannot. Take it, please."

"Thank you, but –"

"I must insist. As one scientist to another. This collection serves no purpose locked away in my study. Take them and show them to people in your country who know nothing about the islands of the Bering Sea."

George saw that to refuse any further would be impolite. "Mr. Grebnitzky," he said. "Your generosity will not go unacknowledged. Thank you."

"It is my deepest pleasure to develop understanding between our nations," said Nicolai Grebnitzky, extending his hand to George.

As the commissioners prepared to return to the *Danube*, Mr. Grebnitzky appeared with two small crates. "In these crates," he said, "you will find the complete skeletons of two fur seals, one cow and one pup. All the bones have been numbered and marked for reconstruction."

⌒

As the *Danube* steamed west to the Kamchatka Peninsula, the volcanic peaks of Mts. Kronotzki and Japorouski rose out of a blue fog and floated against a clear sky. Steering south to Petropavlovsk, George remained on deck to watch a lucid sunset of purple and pink, and a sickle moon that drifted down the inky sky as it followed the sun towards tomorrow.

Petropavlovsk, in daylight, was a ruinous-looking place. The earthworks of the *barabaras* were decayed, the potato gardens were neglected, and the old orchards stood in gnarled rows like weary skeletons. George purchased a reindeer skull and antlers, and a robe of red fox pelts for twenty dollars.

By now it was mid-September, and the weather could deteriorate rapidly at any time, so the captain began the return voyage immediately, steaming directly for the Aleutians for a final inspection of the rookeries before following the continental coast back to Vancouver.

During the night the ship began to plunge and roll in heavy beam seas. Drawers in George's cabin flew open, spilled their contents on the floor and slammed shut with each roll of the ship. Sleep was impossible, so George braced himself in his bunk and fastened the catch cloth to prevent himself from being tossed onto the floor.

George and Sir George were the only ones to attend breakfast, a light affair because of the difficulty of keeping anything on the galley stoves or on the tables in the saloon, and lunch was a sandwich taken standing in the hallway while holding on with one hand. Even reading and writing were difficult, making for a tedious and unproductive crossing.

After taking on coal and water at Unalaska Island, the *Danube* carried on down the coast, making stops to interview natives and whites at Shakan, Port Simpson, Metlakatla, Port Essington, Massett, Nahwitti, Clayoquot Sound, Barkley Sound, and Neah Bay, Washington, before arriving in Vancouver on the 18th of October, 1891.

Sir George Baden-Powell was by now quite impatient to be headed back East to greater comforts. The very next day they boarded the train for the East, stopping only once, in Canmore, Alberta, so he could inspect a mine in which he owned shares.

George remained in Ottawa while Sir George returned home to England. In January of 1892, the two men met again and went to Washington, D.C., for meetings with the American committee. Then, in April, they travelled to England to prepare for the arbitration hearings to be held in Paris, in September.

"Skookum Tumtum was not only a strong enduring man, but a cheery, brave man."
G.M. Dawson with crew at Fort McLeod, 14 July 1879.

Epilogue

> The acid test of any geologist is how well his
> work will stand up when it is re-examined
> later on the ground by a geologist equipped
> with new tools and informed by new scien-
> tific knowledge. Dawson meets this test bet-
> ter than any man of his generation.
>   – Zaslow, *Reading the Rocks*

The Bering Sea Arbitration in Paris found that
Canadian sealing vessels had been seized unlaw-
fully by the Americans, in violation of the principle of
freedom of the high seas. As a result of his work on
behalf of his country, George was awarded the
Companion of St. Michael and St. George, presented
to him by Queen Victoria.

Lord Alverstone, lord chief justice of England,
said of George's contributions: "It is not possible to
overrate the service which Dr. Dawson rendered us. I
consulted him throughout on many questions of difficulty and never found his judgment to fail, and he was
one of the most unselfish and charming characters I
have ever met."

∞

George received many honours and degrees through-out his career, including: D.Sc. from Princeton, LL.D.s from Queen's, McGill, and Toronto. In 1891 he was elected a Fellow of the Royal Society and two years later elected President of the Royal Society of Canada. In 1896 he became President of the Geological Society of America, and in 1897 he was President of the geo-logical section of the British Association for the Advancement of Science. Also, in that year, George was awarded the Gold Medal of the Royal Geographical Society.

In 1895, with the retirement of Dr. Selwyn, George was made Director of the Geological Survey of Canada. This meant an increase in administrative duties and political enterprise, but George worked hard to keep his summers in the field as it became increasingly difficult to leave Ottawa for such long periods.

On Thursday, February 28, 1901, George spent the day at his office in Ottawa, writing letters in an effort to gain approval for a new building for the GSC. He wanted a new building to house and display the considerable collections of the Survey, as the current building had been inadequate from the very beginning. After work, he dined at the Rideau Club, and went home with a bit of a cough.

The next day, Friday, George missed his only day of work in his entire career. Overnight, he had devel-oped acute bronchitis. "Send a telegram to mother," he instructed his landlady. "Tell her to come at once."

When George's mother arrived at his room at Victoria Chambers on Saturday evening, the 2nd of March, she was informed that George had breathed his last breath only fifteen minutes earlier. He was fifty-two years old. On a scrap of paper by his bed, she found George's handwriting: "Life's good night is God's good morrow to eternal light."

George Dawson's death was widely mourned. Tributes and eulogies appeared in newspapers and scientific journals in Canada, the U.S., and England, all filled with the highest of praise.

Perhaps the most telling praise, though, came from the men he had worked alongside in Canada's northwest. As one friend wrote: "It was at night round the camp fires that he opened up, it was a treat to listen to him." Another wrote in B.C.'s *Mining Journal*:

His readiness to share all work, and laugh at every hardship, was the reason for his extraordinary popularity with the Indians, who are not generally eloquent in their praise of white men. I have it from the lips of Indians, that the Doctor was not only "Skookum," but had a "Skookum Tumtum"; to translate, "Was not only a strong enduring man, but a cheery, brave man, ready to endure all things and suffer all things, saying nothing, or making a merry jest of what some travellers might call dangerous hardships."

In 1927, a big game hunter from New York, Mr. Fenley Hunter, lugged a concrete post to the rim of the Liard River Canyon and planted it along the B.C.–Yukon border.

George's annual Geological Survey Reports had inspired Fenley Hunter to travel the North and attempt to duplicate what was perhaps George's finest season of travel. In three months, in 1887, George Dawson had travelled on the Stikine, Dease, Liard, Frances, Pelly, and Yukon rivers, a voyage of nearly three thousand kilometres. Fenley Hunter, with two good men and a motor for his canoe, had been unable to equal Dawson's travel time on the same route.

The post erected by Fenley Hunter was engraved:

IN MEMORY OF
GEORGE MERCER DAWSON
GEOLOGIST SCIENTIST EXPLORER

Born Aug. 1, 1849
Died March 2, 1901
This monument marks the site of the original wooden post erected by Dawson in 1887 as a reference mark for the B.C.–Yukon Boundary.

Latitude 60 deg. 1'6"

George Mercer Dawson, May 1885

Chronology of
George Mercer Dawson
(1849-1901)

Compiled by Lynne Bowen

DAWSON AND HIS TIMES	CANADA AND THE WORLD
	1720 Benjamin Martin speculates that tuberculosis (TB) is caused by tiny living creatures.
	1793 Explorer Alexander Mackenzie reaches the mouth of the Bella Coola River and sees the Pacific Ocean.
	1798 Smallpox vaccine is introduced into North America.
	1808 While looking for the mouth of the Columbia River, explorer Simon Fraser, discovers a new river which explorer and map maker David Thompson will name after

George Mercer Dawson

Fraser, who has already named the Thompson River after him.

1811
James Dawson (Dawson's paternal grandfather), a stern Presbyterian, arrives in Nova Scotia from Banffshire, Scotland.

1818
Convention of 1818 agrees that the forty-ninth parallel will be the border between British North America (B.N.A.) and the United States of America (U.S.) from Lake of the Woods to the summit of the Rocky Mountains.

1820
(John) William Dawson (Dawson's father) is born at Pictou, Nova Scotia; his father, James, is now a printer and bookseller.

1820
Explorer Sir Alexander Mackenzie dies in Britain.

1824
David Thompson surveys the boundary between the U.S. and Canada from the St. Lawrence River to Lake of the Woods.

The Geological Survey of Canada (GSC) is founded to assist in developing a mineral industry; it begins its work in a warehouse in Montreal.

Alfred Richard Cecil Selwyn (future director of the GSC) is born in Kilmington, England.

DAWSON AND HIS TIMES	CANADA AND THE WORLD
	1827 Sandford Fleming (future Canadian railway surveyor) is born in Scotland.
	1830 Charles Lyell, Scottish geologist, publishes the first of his three-volume *Principles of Geology;* the textbook influences many, including naturalist Charles Darwin.
	1831 Charles Darwin sails on H.M.S. *Beagle* for a five-year-long scientific voyage.
	1833 William Fraser Tolmie, Scottish surgeon, comes to Canada to work for the Hudson's Bay Company (HBC).
1840 William Dawson goes to Edinburgh University to study geology; he meets and courts Margaret Mercer.	
1841 Restricted family finances force William Dawson to return to Nova Scotia.	**1841** In London, England, the Royal School of Mines, an auxiliary of the Geological Survey, enrolls its first students.
1842 Charles Lyell and William Dawson spend four weeks together establishing the study of Nova Scotia geology.	

DAWSON AND HIS TIMES	CANADA AND THE WORLD
1846	**1846**
William Dawson returns to Edinburgh and a year later marries Margaret Mercer despite her family's opposition; the couple returns to Nova Scotia.	Oregon Treaty sets the forty-ninth parallel as the border between B.N.A. and the U.S. from the summit of the Rocky Mountains west to the Strait of Georgia.
1848	**1848**
Margaret Dawson gives birth to her first son, James.	Geologist Charles Lyell is knighted.
	California gold rush begins.
1849	**1849**
Infant James Dawson dies in July.	A Coast Salish man reports the presence of coal at Nanaimo on Vancouver Island to the HBC in Victoria.
George Mercer Dawson is born on August 1, to William and Margaret Dawson in Pictou, Nova Scotia.	
William Dawson begins lecturing at Dalhousie College.	In England, Charles Dickens publishes the first installment of *David Copperfield*.
1850	**1850**
William Dawson becomes Nova Scotia's first Superintendent of Education.	Joseph McKay, a HBC clerk, observes coal in Nanaimo on Vancouver Island.
1851	**1851**
Anna Lois Dawson (Dawson's eldest sister) is born.	The threepenny beaver, Canada's first postage stamp, designed by Sandford Fleming, is issued.
	Englishman F. Scott Archer invents the wet collodion process, which makes it possible for photographers to produce clear glass negatives and thus duplicate photographs.

1852
On Sir Charles Lyell's second visit to Nova Scotia, he and William Dawson make important geological discoveries.

1852
Charles Dickens publishes *Bleak House*.

1853
Vaccination against smallpox is made compulsory in Britain.

1854
William Bell Dawson (Dawson's brother) is born.

1854
Britain, France, and Turkey declare war on Russia in the Crimea; Alfred, Lord Tennyson publishes the poem "The Charge of the Light Brigade."

1855
The Dawson family moves to Montreal when Dawson's father becomes principal of McGill College and publishes *Acadian Geology*.

1855
Pend Orielle River gold rush begins in Washington Territory.

1856
Sir Charles Lyell urges Charles Darwin to write about his theories of evolution by natural selection.

1857
John Palliser begins a three-year-long exploration of the western plains of B.N.A. along the unsurveyed American boundary.

c.1858
In the winter, Dawson begins to feel unwell with headaches and body pain.

1858
In British Columbia (B.C.), the Fraser River gold rush attracts large numbers of men, many of them Chinese; the mainland colony of B.C. is established.

George Mercer Dawson

DAWSON AND HIS TIMES	CANADA AND THE WORLD
Ottawa is chosen to be the capital of Canada.	In the German province of Westphalia, future American anthropologist, Franz Boas, is born.

c.1859

By March Dawson's health seems better; but during the following winter the pain returns to his back and neck; he is bedridden and his back becomes twisted and hunched.

1859

Charles Darwin publishes *On the Origin of Species by Means of Natural Selection*.

Tennyson publishes *Idylls of the King*.

c.1860

Dawson is sent to Pictou to live by the sea with his grandfather; his condition continues to deteriorate; his father sends him to Dr. Buckminster Brown near Boston, who diagnoses Pott's disease or TB of the spine; Dawson is placed in a body truss to prevent further distortion of his upper body.

1860

In B.C., prospectors discover gold on the Horsefly River in the Cariboo Mountains.

Humphrey Hime, Canadian photographer and surveyor, publishes a portfolio of prints taken during the Saskatchewan and Assiniboine exploring expedition of 1858.

1861-1865

Dawson is bedridden for several years before his strength begins to return; he receives school lessons at home and reads extensively (Dickens, Byron, Tennyson); the family spends the summer at Tadoussac.

Anglican lay preacher, William Duncan, moves with his native converts to the new settlement of Metlakatla on the B.C. North Coast, where he hopes the native people will abandon their traditional ways.

1861

British Boundary Commission completes its survey of the forty-ninth parallel from the Pacific coast to the summit of the Rocky Mountains.

The American Civil War begins.

Pauline Johnson, future poet, is born near Brantford, Ontario.

1862

Smallpox epidemic kills one third of the native population of B.C.

Prospectors find gold near bedrock at Barkerville in the Cariboo region of B.C.

Dawson and His Times	Canada and the World
1863	**1863**
Rankine Dawson (Dawson's youngest brother) is born.	GSC publishes its first comprehensive report, *Geology of Canada*.
	In B.C., news of Cariboo gold brings miners and adventurers from all over the world.
1864	**1864**
Eva Dawson (Dawson's youngest sister) is born.	In B.C., the Cariboo Road opens to provide a wagon route to the goldfields at Barkerville; several labourers building a wagon road from Bute Inlet to the Interior are murdered by the Tsilhqot'in (Chilcotin) five of whom are tried and executed.
1865	**1865**
Dawson accompanies his father and sister, Anna, to Britain where he stays with relatives in Scotland; back in Montreal his interests are those of a naturalist.	In the U.S., Abraham Lincoln is assassinated; the Civil War ends.
1866	**1866**
Dawson's father gives him a small rifle, which he learns to shoot.	B.C. and Vancouver Island colonies unite.
	1867
	BNA Act establishes the Dominion of Canada uniting Ontario, Quebec, Nova Scotia, and New Brunswick.
	U.S. purchases Alaska from Russia and the dispute between Canada and the U.S. over the boundary of the Alaska Panhandle begins to smoulder.

DAWSON AND HIS TIMES	CANADA AND THE WORLD
1868 Dawson enrolls at McGill College.	**1868** In the U.S., William (Buffalo Bill) Cody is employed as a hunter to supply buffalo meat to the Kansas Pacific Railroad.
1869 Dawson sails for Glasgow aboard the *Lake Erie*; he stays with relatives and then goes to London, where he enrolls in the Royal School of Mines and lodges at 20 Halsey Street in Chelsea.	**1869** A.R.C. Selwyn succeeds Sir William Logan as the director of the GSC and assumes the enormous task of directing the examination of the vast new territories in the west, which are about to become part of Canada. In the U.S., the first transcontinental railroad and telegraph is completed; it improves transportation and communication dramatically for the entire West Coast. Robert Dunsmuir discovers the Wellington coal seam on Vancouver Island. In Manitoba, Louis Riel leads the Red River Rebellion.
1870 Having finished the year without distinguishing himself at school, Dawson goes to Edinburgh to spend the summer with his parents.	**1870** HBC sells Rupert's Land to the government of Canada; the sale extends Canadian territory from the Ontario-Manitoba border to the Pacific Ocean; the area to be covered by the GSC is increased tenfold; the GSC shifts its emphasis to railway land grants; Manitoba joins Canadian Confederation.

1871

Dawson has acquired some friends and improved his marks; he wins the Duke of Cornwall's Scholarship, the Director's Medal and Prize, and the Edward Forbes Medal and Prize; he works in the Lake District of England for part of the summer.

1872

Dawson finishes first in his class and wins the Forbes competition a second time; he graduates as an Associate of the Royal School of Mines.

1873

Dawson goes to Quebec City in January to teach chemistry.

William Dawson uses his influence with his friends Samuel Tilley and Joseph Howe to secure for his son the post of naturalist and geologist with the joint British and

1871

B.C. enters Canadian Confederation on the promise that a transcontinental railway will be finished within ten years.

The Canadian Pacific Railway (CPR) appoints Sandford Fleming engineer in charge of the major surveys across the prairies and through the Rocky Mountains.

A.R.C. Selwyn surveys southern B.C.; James Richardson begins the first of four consecutive summers studying Vancouver Island coal deposits and Queen Charlotte Island (Haida Gwaii) anthracite beds.

Emily Carr, future painter and writer, is born in Victoria, B.C.

1872

In September, the Joint Boundary Commission assembles in Pembina, North Dakota; in November the Americans go home and the British and Canadians settle at Fort Dufferin, Manitoba, for the winter.

1873

Samuel Tilley, politician and member of Canada's first cabinet, is appointed Lieutenant-Governor of New Brunswick.

Joseph Howe, politician and opponent of Canadian Confederation, becomes Lieutenant-Governor of

American survey of the international boundary.

Nova Scotia three weeks before he dies.

In June, Dawson travels by steamer across Lake Superior to Duluth, Minnesota, by rail to Moorhead, North Dakota, then up the Red River into Canada by riverboat and wagon to Fort Dufferin, Manitoba.

Prince Edward Island joins Canadian Confederation.

Canadian Parliament creates the North-West Mounted Police.

Dawson travels 459 kilometres west of Red River to Porcupine Creek during his first summer with the survey; he and his horse are swarmed by locusts.

1874

Returning to Montreal, Dawson spends the winter identifying specimens and writing field notes.

Dawson rejoins the Boundary Commission and they begin work again at Porcupine Creek; they visit the Métis camp in the Cypress Hills where the last generation of buffalo hunters live; by August the survey reaches the Rockies at Waterton Lake; they return to Fort Dufferin having travelled 1384 kilometres overland.

Dawson returns to Montreal and publishes his "Report on the Geology and Resources of the Region in the Vicinity of the Forty-Ninth Parallel"; still in agreement with his father's position, Dawson states his belief that glaciers did not cover the prairies,

1874

George Anthony Walkem becomes premier of B.C. for the first of two terms which will be dominated by the struggle with Ottawa over its failure to begin the transcontinental railway in the time specified when B.C. joined Confederation.

Indians and Chinese lose the right to vote in B.C. elections.

but rather floated west when the continent was covered by water; "erratics" were deposited as the water receded.

1875
On July 1, Dawson is appointed a geologist with the GSC; he travels by railway to San Francisco and steamer to Victoria, B.C.; most of the province is untouched by geological exploration; he spends the summer and fall with CPR survey parties exploring the central Interior west of the Fraser River and looking for possible railway routes and the agricultural and resource potential of the area; he returns to Vancouver Island and inspects the Douglas Mine at Nanaimo and its coal seam.

1875
A.R.C. Selwyn examines the Peace River District as a possible northern route for the CPR.

In Britain, Sir Charles Lyell dies and is buried in Westminster Abbey.

1876
In Victoria, Dawson works with Dr. Tolmie during the winter to compile vocabularies on First Nations languages and dialects.

In May, Dawson heads for the Interior of B.C. and meets with Dan McMillan, an engineer for the CPR survey; they spend the summer exploring the central Interior west of the Fraser River.

In the fall Dawson returns to Montreal.

1876
American cavalrymen under Lt. Col. George Custer are massacred by the Sioux and Cheyenne at the Little Big Horn River.

Alexander Graham Bell invents the telephone.

1877
Dawson spends his summer in the Thompson-Shuswap region and in

1877
The surrender of Sioux chief Crazy Horse marks the end of the

the Fraser, Similkameen, Okanagan, and Nicola Valleys; he always looks for evidence of glaciation and often finds it.

Dawson steals Aboriginal skulls from a burial site at Lytton, B.C.

Dawson's inventory of B.C. ore deposits is published; he receives an honourary degree from Princeton University.

American Plains Indian wars; Sioux Chief Sitting Bull escapes to Canada.

Queen Victoria is proclaimed the Empress of India.

1878
Dawson spends the summer examining northern Vancouver Island and the Queen Charlotte Islands; he names many geographical features; as in all his field work, he must serve as a mapmaker too because no maps of B.C. have previously been made; his reports each year also include meteorological information; he photographs Haida villages.

1878
Paris hosts a World Exhibition.

1879
Dawson's report on the species of harvestable trees is published.

As he finds more and more evidence supporting glaciation, Dawson abandons his drift-ice hypothesis for the Prairies and B.C.

1880
Dawson publishes an ethnological study of the Haida nation.

1880
Sandford Fleming retires from the CPR but continues to act as a consultant; he advocates the adoption of standard time and time zones.

DAWSON AND HIS TIMES	CANADA AND THE WORLD
Dawson spends his summer in Montreal, where he avoids people, as is his custom, but finds the solitude of the city unnatural and oppressive.	In Canada and the U.S., the prairie bison (buffalo) have been hunted almost to extinction. Thomas Edison and J.W. Swan each devise the first practical electric lights.
1881 Dawson advocates integration of Indians into Canadian society and says that they possess the qualities that would make this possible; he is, however, disdainful of their way of living and describes the potlatch as "pernicious."	**1881** CPR is incorporated; the Kicking Horse Pass is chosen as the railway's route through the Rocky Mountains; large numbers of Chinese labourers are imported to build the railway through the mountains.
1882 Dawson is a charter member of the Royal Society of Canada, which has been formed by his father. In an article for *Harper's New Monthly Magazine* Dawson states that the original inhabitants of North America probably came from Eastern Asia.	**1882** Charles Darwin dies in England. John Macoun, explorer and naturalist, is appointed to the GSC as Dominion botanist and is a charter member of the Royal Society of Canada. Robert Koch discovers the tubercle bacillus and demonstrates its role in causing TB.
1883 Dawson becomes Assistant Director of the GSC.	**1883** Buffalo Bill Cody organizes his "Wild West Show," which will tour Europe and the U.S. until 1916.
1884 Dawson and Selwyn publish *Descriptive Sketch of the Physical Geography and Geology of the Dominion of Canada*.	**1884** In Washington, D.C., Sandford Fleming convenes the International Prime Meridian Conference; international standard time is adopted.

DAWSON AND HIS TIMES	CANADA AND THE WORLD

Dawson's father is knighted for his public services.

Tolmie and Dawson publish *Comparative Vocabularies of the Indian Tribes of British Columbia.*

Dawson is a member of the Committee on the Northwestern Tribes and later arranges funds so that most of the artifacts collected by Franz Boas remain in Canada.

In Canada, the federal government bans the Northwest Coast potlatch; the ban will last for sixty-six years.

British Association for the Advancement of Science holds its first Canadian meeting and establishes the Committee on the North-western Tribes which engages Franz Boas to make a series of field expeditions to B.C.

1885
Dawson conducts a survey of Vancouver Island; in keeping with his desire to preserve a record of native life, he takes a large collection of Kwakiutl artifacts to the GSC Museum.

1885
Louis Riel returns from exile to lead the Métis in the Northwest Rebellion in present-day Saskatchewan; although the railway is unfinished, the CPR makes it possible for troops to travel from Ontario to the battlefields.

A major outbreak of smallpox in Montreal results in over 3000 deaths and causes public health authorities to attempt to enforce compulsory vaccination, which is resisted by the populace; this plus the news that Louis Riel has been sentenced to death causes street rioting in Montreal.

The Last Spike of the CPR is driven at Craigellachie in B.C.'s Eagle Pass in November.

Chinese are denied the vote in Canadian federal elections and new arrivals must pay a fifty-dollar head tax.

DAWSON AND HIS TIMES	CANADA AND THE WORLD

1886
The first CPR passenger train leaves Montreal on June 28 and arrives at Port Moody, B.C. on July 4.

Franz Boas shifts his field research permanently to the Northwest Coast.

1887
Beginning what would later be called "his finest season of travel," Dawson explores the south and central Yukon territory and investigates gold discoveries there; in September, Dawson and his crew cross the Chilkoot Pass having circumscribed an area of over 100,000 square kilometres.

Dawson publishes an ethnological study of the Kwakiutl nation.

1887
The British Empire celebrates Queen Victoria's Golden Jubilee.

1889
Dawson's revised inventory of B.C. ore deposits is issued.

1889
North and South Dakota, Montana, and Washington become states; Oklahoma is opened to non-Indian settlement.

1890
Dawson states his final position on glaciation: a huge ice mass, the Cordilleran Glacier, covered B.C.

1890
Norman Bethune is born in Gravenhurst, Ontario.

1891
Dawson is elected a Fellow of the Royal Society.

Dawson and British stateman Sir George Baden-Powell are

1891
British government forms the Bering Sea Commission to respond to American threats to the fur seal fishery off the coast of Alaska.

George Mercer Dawson

appointed to the British Bering Sea Commission; they travel to the Bering Sea and meet with Russian and American delegates.

Canada's first prime minister, Sir John A. Macdonald, dies in Ottawa.

Dawson and Baden-Powell return to Vancouver in October and from there they go to eastern Canada by train.

Dawson publishes an ethnological study on the Shuswap nation.

1892
In January, Dawson and Baden-Powell go to Washington, D.C. to meet with the American committee; in April they go to England to prepare for the arbitration hearings to be held in Paris in September; the Bering Sea Arbitration finds the U.S. guilty of violating the principle of freedom of the high seas; Dawson is awarded the Companion of St. Michael and St. George by Queen Victoria.

1892
Poet Pauline Johnson performs for the first time in public at the Academy of Music in Toronto.

1893
Dawson is elected President of the Royal Society of Canada.

William Dawson publishes *The Canadian Ice Age*, which also contains extracts from his son's publications.

1893
In Germany, Karl Benz constructs his four-wheel car; in the U.S., Henry Ford builds his first car.

Geologist, Scientist, Explorer

DAWSON AND HIS TIMES	CANADA AND THE WORLD
	1894 Poet Pauline Johnson tours western Canada and travels to B.C. by train. Robert Service, future "Poet of the Yukon," immigrates to Canada.
1895 Dawson is made Director of the GSC, which means living in Ottawa and spending more time in administrative and political duties; he continues to spend his summers in the field.	**1895** A.R.C. Selwyn retires as director of the GSC. In France, the Lumière brothers patent their motion picture camera.
1896 Dawson becomes President of the Geological Society of America and a member of the Ethnological Survey of Canada.	**1896** Placer gold is discovered on Bonanza Creek in the Klondike in Yukon territory. Wilfrid Laurier becomes Canada's first prime minister of French ancestry.
1897 Dawson becomes President of the geological section of the British Association for the Advancement of Science; he is awarded the Gold Medal of the Royal Geographical Society.	**1897** Canada's Sandford Fleming is knighted. The British Empire celebrates Queen Victoria's Diamond Jubilee. Word of Klondike gold reaches the world, and a stampede of prospectors begins by water and land.
1898 Dawson City is named after George Dawson.	**1898** First photographs using artificial light are taken.

163

George Mercer Dawson

DAWSON AND HIS TIMES	CANADA AND THE WORLD

1899

Dawson's father, Sir John William Dawson, dies in Montreal; the life-long devout Christian is called "the leading anti-Darwinist of the late Victorian period."

1899

Boer War begins in South Africa; Canada sends troops; the war divides Canadians along French-English lines.

Joint High Commission fails in its attempt to solve the Alaska Boundary dispute between the U.S. and Canada, which has come to a head since the Klondike gold rush began.

1900

Mortality rate for TB in Canada is 180 per 100,000 and climbing.

1901

On February 28, George goes home with a cough; the next day he has acute bronchitis and misses the only day of work in his career; he sends a telegram to his mother telling her to come at once; when his mother arrives on March 2, George has just died.

1901

Queen Victoria dies and is succeeded by her son, King Edward VII.

Marconi transmits telegraphic radio messages from Cornwall to Newfoundland.

1902

A.R.C. Selwyn dies at Vancouver, B.C.

1903

Canada feels betrayed when the British agree with the Americans in the settlement of the Alaskan Boundary Dispute.

1927

American Fenley Hunter installs a concrete post at the rim of the Liard River Canyon at the B.C.-Yukon border on the site of

164

DAWSON AND HIS TIMES

CANADA AND THE WORLD

Dawson's original wooden marker,
installed in 1887.

Sources Consulted

BARKHOUSE, Joyce. *George Dawson: the Little Giant.* Toronto: Natural Heritage, 1989.

CANNINGS, Sydney, and CANNINGS, Richard. *Geology of British Columbia: A Journey Through Time.* Vancouver: Douglas & McIntyre, 1999.

CLASSEN, George C. *Thrust and Counter-Thrust: The Genesis of the Canada–U.S. Boundary.* New York: Rand McNally & Co., 1967.

COLE, Douglas, and LOCKNER, Bradley, eds. *The Journals of George M. Dawson: British Columbia, 1875-1878,* 2 vols. Vancouver: UBC Press, 1989.

COLE, Douglas, and LOCKNER, Bradley, eds. *To The Charlottes: George Dawson's 1878 Survey of the Queen Charlotte Islands.* Vancouver: UBC Press, 1993.

DALZELL, Kathleen E. *The Queen Charlotte Islands, Volume 2, Places and Names.* Queen Charlotte City: Bill Ellis, Publisher, 1981.

DAWSON, Dr. George Mercer. "Geological notes on some of the coasts and islands of the Bering Sea and vicinity," 1892.

DAWSON, Dr. George Mercer. "Historical notes on the Yukon district."

DAWSON, Dr. George Mercer. "Notes on the Locust Invasion of 1874 in Manitoba and NWT."

DAWSON, Dr. George Mercer. Private Diaries, 1887: vol. I & II.

DAWSON, Dr. George Mercer. Private Diaries, 1891: vol. I & II.

DAWSON, Dr. George Mercer. "Report of the Bering Sea Commission."

DAWSON, Dr. George Mercer. "Report on an exploration in the Yukon district, NWT, and adjacent northern portions of B.C.," 1887.

DAWSON, Dr. George Mercer. "Report on the Geology and Resources of the Region in the Vicinity of the Forty-Ninth Parallel," 1874.

DAWSON, Dr. George Mercer. "The Yukon Territory: the narrative of an exploration made in 1887 in the Yukon District."

DAWSON, Dr. George Mercer. "The Superficial Geology of British Columbia and Adjacent Regions," in *Quarterly Journal of the Geological Society*. May 1881.

FITZGERALD, William G. "The Romance of Seal Hunting: An Interview with Sir George Baden-Powell, K.C.M.G., Etc." *The World Wide Magazine*, Vol. 1. No. 1. April, 1898.

GAYTON, Don. *The Wheatgrass Mechanism: Science and Imagination in the Western Canadian Landscape*. Saskatoon: Fifth House Publishers, 1990.

MACASKIE, Ian. *The Long Beaches: a Voyage in Search of the North Pacific Fur Seal*. Victoria: Sono Nis Press, 1979.

ROED, Murray A. *Geology of the Kelowna Area and the Origin of the Okanagan Valley, British Columbia.* Kelowna: Kelowna Geology Committee, 1995.

STEGNER, Wallace. *Wolf Willow.* New York: Viking Press, 1971.

TIPPER, H.W. *Glacial Morphology.* GSC Bulletin no. 196. 1977.

THOMPSON, Donald W. *Men and Meridians.* Vol. 2, 1867-1917. Ottawa: Queen's Printer, 1972.

WINSLOW-SPRAGGE, Lois. LOCKNER, Bradley, ed. *No Ordinary Man: George Mercer Dawson.* Toronto: Natural Heritage, 1993.

WINSLOW-SPRAGGE, Lois. *The Life of George Mercer Dawson: 1849-1901.* Montreal: 1962.

ZASLOW, Morris. *Reading the Rocks: the Story of the Geological Survey of Canada 1842-1972.* Toronto: Macmillan Company of Canada, 1975.

Index

Aboriginal peoples, 11, 24, 50, 72-74, 84, 114, 115, 119-20, 124, 125-26, 138, 156, 159
and Christianity, 53, 55, 152
legends of, 57-58, 69-71, 73-79, 105-06
See also Aleuts; Assiniboines; Bella Bellas; Sioux; Gros Ventre; Blackfeet; Chilcotins; Crow; Cree; Haida; Interior Carrier; Interior Salish; Klusklus; Kwakiutl; Pelly; Tlingits; Tsimshians
Akutan Pass (Alaska), 128
Alaska, 127, 129, 153, 161
Alaska Commercial Company, 129, 130
Alaskan, 107, 109-10
Alaskan Boundary Dispute. *See* Alaska panhandle
Alaska panhandle, 153, 164
Albatross, 129
Aleutian Islands, 126, 127, 129, 138
Aleuts, 131, 136
Alexis Creek (B.C.), 52
Alliford Bay (B.C.), 101
Allison Pass (B.C.), 81
Alverstone, Lord (lord chief justice of England), 141
Americans, 4, 5, 9
Anahim Lake (B.C.), 71
"Angle." *See* Northwest Angle
Anthony Island (B.C.), 97-98
Ashcroft Manor, B.C., 50
Assiniboine, District of, 152
Assiniboines, 16, 23

Baden-Powell, Sir George (Bering Sea Commissioner), 126, 127, 131, 132, 138-39, 161-62
Barkley Sound (B.C.), 138
Bella Bella, B.C., 95, 105
Bella Bellas, 68
Bella Coola River (B.C.), 147
Bella Coola Trail, 69
Bennett Lake (Yukon), 123
Bering Island, 135, 137
Bering Sea, 124, 126-27, 128, 137, 162
Bering Sea Arbitration. *See* Bering Sea Commission
Bering Sea Commission, 127, 132, 135, 139, 141, 161, 162
Bering, Vitus (explorer), 129
Bethune, Norman (physician), 161
Big Camp, 21-22, 25
Blackfeet, 23, 25, 28
See also Dream beds
Black flies, 68
Blackwater River, 56, 59, 64, 66, 67
Boats. *See* Canoes and boats
Boas, Franz (anthropologist), 152, 160, 161
Boer War, 164
Bonanza Creek. *See* Klondike gold rush
Boundaries, 144, 153
See also Forty-ninth parallel
Bowman, Amos (Dawson's assistant), 64, 68-69, 72
British-American Boundary Survey. *See* British North American Boundary Survey Commission

British Association for the Advancement of Science, 142, 163
British Boundary Commission, 30
British Columbia (B.C.), 5, 7, 8, 22, 30, 31, 48, 113, 151, 153, 155, 156, 157, 158, 160, 161, 163, 164
British North America Act (BNA), 153
British North American Boundary Survey Commission, 2, 4-5, 6, 7-14, 15, 20-21, 22, 155, 156
Brown, Dr. Buckminster (Dawson's doctor), 39-40, 152
Bruin Bay (B.C.), 105
Brulé, 56, 64, 66
Buffalo, 11, 16, 21-22, 23-24, 26-28, 154, 156, 159
Bute Inlet (B.C.), 153
Byron, Lord (poet), 152

Camels, 82-84
Campbell, Mr. (messenger), 111-12
Campbell, Robert (map maker), 112, 115, 120
Canada-United States Boundary. *See* Forty-ninth parallel
Canadian Boundary Survey Commission. *See* British North American Boundary Survey Commission
Canadian Confederation, 153, 154, 155, 156
Canadian Pacific Railway (CPR), 54, 64, 68, 156, 159, 160, 161
survey, 56, 155, 157, 158
Canadian Shield, 23
Canmore, Alberta, 138
Canoes and boats, 33-36, 38, 58-59, 84-88, 98, 99, 105, 111, 113, 116, 119, 121, 122, 125-26, 131

Cape Fanny (B.C.), 97
Cape Kalekhta, Alaska, 128
Cape Knox (B.C.), 105
Cape St. James (B.C.), 97
Captain Jim (Tsimshian boatman), 113, 115-17
Captain Jim's wife, 113, 114, 116
Cariboo region (B.C.), 152, 153
Carpenter Bay (B.C.), 98, 99
Carpenter, William Benjamin (English naturalist), 98
Carr, Emily (painter), 155
Cassiar, B.C., 109
Charley (Aboriginal man from Fraser Lake), 72-79
Charley (Chilcotin man), 57-58
Charley (Tsimshian boatman), 113, 114, 115, 116-17
Cheddakulk Creek (B.C.), 71
Chief Klue, 100-01, 103
Chilcotins, 52-53, 57-58, 153
Chilcotin River (B.C.), 52, 53, 57
Chilkoot Pass (Yukon), 121, 123
Chinese, 109, 151, 156, 159, 160
Chinook Jargon, 52, 57, 65, 73, 96, 100, 102, 106, 110, 119
Clayoquot Sound (B.C.), 138
Clinton, B.C., 50
Coal deposits and workings, 71, 80, 97, 101, 150, 154, 155, 157
Coast Range Mountains (B.C.), 52
Commander Islands, 129
Coldwater River (B.C.), 80
Coppermine River, 34
Cordilleran glacier. *See* Glaciation
Coquihalla River (B.C.), 81
Cree, 23
Crimean War, 151
Crow, 23, 24-25
Culla-Culla House (B.C.), 62, 68, 70
Cypress Hills (District of Assiniboia), 21-22, 156

Dakota, 76
Dalhousie College, 37
Danube, 125, 127, 128, 129, 130, 131, 132, 135, 137, 138
Darwin, Charles (scientist), 149, 151, 152, 159, 164
Dawson, Anna Lois (Dawson's sister), 6, 32, 36, 37, 39, 40, 150, 153
Dawson City (Yukon), 123
Dawson, Eva (Dawson's sister), 153
Dawson, George Mercer
  and Aboriginal people, 99-101, 143, 157, 159, 160, 161, 162
  appearance of, 2, 7, 8, 39, 40
  birds seen by, 7, 50, 64, 77, 78, 97, 128, 133
  birth of, 36, 150
  childhood of, 33-45, 151-53
  as a collector of fossils, 55-56, 68-69, 89-90
  as a collector of plant and rock specimens, 3, 8, 9, 13, 14, 21, 30, 37, 45, 48-49, 61, 68, 158
  as a collector of skeletal remains, xi, 90-91, 137, 158
  death of, 143, 164
  education of, 5, 7, 39, 40, 41, 45, 152, 154, 155
  equipment of, 6
  illnesses and injuries of, 7, 8, 36, 38-41, 45, 86, 118, 151, 152, 164
  interest in languages of, 61, 157, 160
  honours given to, 141, 142, 155, 158, 161, 162, 163
  leisure pursuits of, 61, 159
  pack trips of, 51-52, 54-58, 64-72, 79-84, 89-91, 118, 157, 161
  as a photographer, 24-25, 44, 46, 62, 72, 76, 92, 101, 158

  plants and trees seen by, 9, 27, 50, 52, 64, 81, 86, 88, 104, 111, 112, 116, 123, 128, 136-37
  as a poet, 10, 45
  publications of, 31, 156, 158, 159, 160, 162
  river, lake, and ocean voyages of, 58-59, 72-79, 85-88, 95-99, 101, 107-16, 119-20, 122-23, 125-26, 128-29, 129-32, 138, 158
  and sketching, 12, 29, 44, 105
  travels of, 17-20, 47-48, 60-61
Dawson, James (Dawson's paternal grandfather), 36-37, 39, 148, 152
Dawson, James (Dawson's infant brother), 36, 150
Dawson, John William, (Dawson's father), 3, 6, 12, 14, 22, 30, 32, 36, 37, 38, 41, 43-44, 45, 90, 148, 149, 150, 153, 164
  career of, 37, 41, 149, 150, 151, 156, 159, 160, 162
Dawson, Margaret née Mercer (Dawson's mother), 32, 36, 38, 40, 41-42, 43-44, 45, 143-44, 149, 150, 164
Dawson, Rankine (Dawson's brother), 93-95, 96, 97, 98, 99, 100, 101, 102, 104, 153
Dawson, William Bell (Dawson's brother), 32, 38, 151
Dean River (B.C.), 64, 68, 71, 72
Dease Lake (B.C.), 110, 112
Dease River (B.C.), 112, 113, 144
Dickens, Charles, 150, 151, 152
Dogs, 136
Dolomite Narrows (B.C.), 99
Douglas, Mr. (ship's captain), 94, 95
Dream beds, 28-29
Drift-ice hypothesis. *See* Glaciation
Duncan, William (missionary), 152

Dunsmuir, Robert (industrialist), 154

Edinburgh, Scotland, 149, 150, 154
Erratics, 11, 12, 23, 157
Ethnological Survey of Canada, 163
Euchiniko River (B.C.), 64, 66

Finlayson Lake, 116
Finlayson River (Yukon), 115
Fires, 13-14, 27
Fishing, 74, 86, 95-96
   salmon, 52, 85, 87-88, 105-06, 122-23
Fleming, Sandford (railway surveyor), 149, 150, 155, 158, 159, 163
Fort Dufferin, Manitoba, 5, 13, 14, 17, 20, 30, 155, 156
Fort Fraser, B.C., 72
Fort George, B.C., 56, 58, 76
Fort McLeod, District of Alberta, 140
Fortymile (Yukon), 121, 122
Forty-ninth parallel, 4, 8, 21, 30, 31, 148, 150, 151, 152
Fossil Creek (B.C.), 55
Frances Lake, 114, 115
Frances River (Yukon), 114, 144
Francois Lake (B.C.), 72, 73
Fraser Canyon (B.C.), 50, 63
Fraser Lake (B.C.), 73, 74, 75
Fraser River (B.C.), 34, 50-51, 52, 58-59, 63, 87, 90, 147, 157, 158
Fraser, Simon (explorer), 34, 35, 147
Frederick Island (B.C.), 105
Frenchman's Creek. *See* White Mud River
Ft. Lewis, 135

Galbraith, Mr. (man from Wrangell), 111
Gardner Channel (B.C.), 64
Gatcho Lake (B.C.), 62, 68, 69, 70, 72
Geological Society of America, 142, 163
Geological Survey (Great Britain), 149
Geological Survey of Canada (GSC), 5, 15, 22, 31, 47, 72, 107, 120, 127, 142, 148, 153, 154, 157, 159, 160, 163
George (Dawson's crewman), 110
Glacial striations. *See* Glaciation
Glaciation, 11-12, 14, 23-24, 29, 30, 54, 71, 74-76, 80, 81, 86, 135, 156-57, 158, 161
   See also Erratics
Glaciers. *See* Glaciation
Gold and gold seekers, 83, 112, 121, 122, 123, 150, 151, 152, 161, 163
   See also Klondike gold rush
Gooseneck barnacles, 104
Grappler, 105
Great Plains, 5, 7, 8
Grebnitzky, Nicolai (naturalist), 135-37
Gros Ventre, 23
Gulf of Alaska, 126

Haida, 95-96, 98, 99-104, 158
Hain-chit (Bella Bella man), 105-06
Harper, Mr. (man from Stewart River), 111, 120, 121
Hayes Creek (B.C.), 89
Hearne, Samuel (explorer), 34, 35
Hime, Humphrey (photographer), 152
Hoole's Canyon, 119
Hope, B.C., 81

Horses, 1, 3-4, 12, 13-14, 18-20, 26, 28, 29, 42-44, 66, 72, 156
pack, 20, 51-52, 56-57, 68, 80, 83, 89
Houston Stewart Channel (B.C.), 97
Howe, Joseph (politician), 155-56
Hudson's Bay Company, 113, 149, 150, 154
Hunter, Fenley (big game hunter), 144, 164
Huxley Island (B.C.), 99

Iliuliuk Habour (Alaska), 128
Ingram, Henry (camel owner), 82-84
Insects. *See* Black flies, Mosquitoes, Locusts
Interior Carrier, 68
Interior Salish, 51, 53
Iron deposits, 80

Jacinto (Dawson's packer), 79, 82, 83, 110
Jason (Aboriginal man from Fraser Lake), 73
Jellyfish, 48-49
Jimmy (Dawson's crewman), 51, 53-56
Johnny (Dawson's cook), 77, 79, 84, 87, 88, 89
Johnson, Davey (Dawson's crewman), 110, 113, 115-16, 117, 118, 120
Johnson, Pauline (poet), 152, 162, 163
Joint Boundary Commission. *See* British North American Boundary Survey Commission
Joint High Commission. *See* Alaska Panhandle

Kamchatka Peninsula, 129, 137

Kamloops, B.C., 79, 84
Kwakiutls, 160, 161
Kawliakas, 131, 132
Kerouard Islands (B.C.), 97
Klondike gold rush, 123, 163, 164
Kokanee. *See* Fishing, salmon
Kootenai Pass, District of Alberta, 29
Klusklus, 65, 66

Lake Laberge (Yukon), 122
Lake of the Woods (Ontario), 4, 30, 148
Larry (Tahltan assistant and cook), 110
Lauder Point (B.C.), 105
Laurier, Sir Wilfrid (politician), 163
Lepas Bay (B.C.), 104
Lewis, "Brick" (Dawson's crewman), 110, 113, 115
Liard River (Yukon), 112, 113, 115, 120, 144, 164
Lindeman Lake (Yukon), 123
Locusts, 1, 3-4, 14, 27, 29, 156
London, England, 154
Lower Post, 112, 115, 120
Lyell Island (B.C.), 99
Lyell, Sir Charles (geologist), 149, 150, 151, 157
Lytton, B.C., 50, 60, 61, 90, 158

Macdonald, Sir John A., 162
McEvoy (Dawson's crewman), 110, 113, 114-16, 117, 120
MacFarlane, Douglas (Dawson's assistant), 79, 81, 84-85, 87, 89, 90-91
McGill College, 14, 32, 37, 45, 90, 94, 142, 151, 154
Mackenzie, Alexander, 70, 147, 148

McMillan, Dan (CPR engineer), 63-67, 157
Macoun, John (explorer and naturalist), 127, 131, 159
Makushin (volcano), 128
Manitoba, 9-10, 154
Marine life. *See* Gooseneck barnacles; Jellyfish; Seals and seal fishery; Sea otters
Marmots, 71
Massett, B.C., 138
Mendenhall, Professor T.C. (Bering Sea commissioner), 129
Merriam, Dr. Hart (Bering Sea commissioner), 129, 133
Métis, 4, 7, 21, 23, 27, 28, 156, 160
See also Big Camp
Metlakatla, B.C., 105, 138, 152
Miles Canyon (Yukon), 121, 123
Mining Journal, 143
Missouri Coteau (District of Assiniboia), 20
Montreal, Quebec, 14, 22, 30, 37-38, 76, 77, 105, 148, 151, 153, 156, 157, 159, 160, 161, 164
Mosquitoes, 9-10, 68, 113, 117
Mount Royal (Quebec), 33, 38
Mules, 66-67, 71-72, 83, 110
Murchison Islands (B.C.), 99

Nahwitti, B.C., 138
Nechako River (B.C.), 72
Nanaimo, B.C., 150, 157
New Westminster. B.C., 50, 77
Nicola Valley (B.C.), 158
Noah (Dawson's paddler), 85, 87
North Dakota, 17, 156, 161
Northumberland Strait (Nova Scotia), 36, 37
Northwest Angle, 4
North-West Mounted Police, 156
Northwest Rebellion, 160
Nova Scotia, 37, 148, 149, 150, 151

Ogilvie, Bill (Dawson's colleague), 111-12, 121-22
O'Hara, Dan (Dawson's friend), 33-36, 42-44
Okanagan Lake (B.C.), 77-79, 81-82, 88, 158
Oregon Treaty, 150
Ottawa, Ontario, 139, 142, 152, 162, 163
Oxen, 10

Palliser, John (explorer), 151
Paris, France, 139, 141, 158, 162
Peace River District, 157
Peachland, B.C., 88
Pelly, 107, 111
Pelly River (Yukon), 108, 111-12, 114, 115, 118, 119, 120, 144
Pembina, North Dakota, 4, 155
Petropavlovsk, Russia, 136, 137
Photography, 150, 163
See also Dawson, George Mercer, as a photographer; Hime, Humphrey
Pictou, Nova Scotia, 36, 37, 39, 148, 150, 152
Polar bear, 133, 134-35
Porcupine Creek (District of Assiniboia), 13, 20, 156
Portages. *See* Dawson, George Mercer, river, lake, and ocean voyages of
Port Essington, B.C., 138
Port Moody, B.C., 161
Port Simpson, B.C., 138
Pott's Disease, 39-40, 152
Prairies, 10-11, 17-22, 27, 29, 31, 48, 155, 156, 158
Pribilof, Gerassim (explorer), 129
Pribilofs (The), 129
See also St. George Island; St. Paul Island
Princeton, B.C., 81, 88, 89

Princeton University, 142, 158
Pseudomorph. See Canoes and
boats

Quebec City, Quebec, 155
Queen Charlotte Islands (B.C.),
94, 96-105, 126, 155
Queen's University, 142
Queen Victoria, 141, 158, 162, 163,
164
Quesnel, B.C., 56, 58, 64, 66, 76

Railways, 154, 155
See also Canadian Pacific
Railway
Ramsay Islands (B.C.), 99
Rankine Islands (B.C.), 99
Redpath Museum, Montreal, 90
Red River (North Dakota and
Manitoba), 4, 6, 8, 13, 156
Red River carts, 21
Reeves (Dawson's cook), 51, 52,
53, 57, 58, 59
Richardson, James (geologist), 48,
155
Riel, Louis (rebel), 154, 160
Riske Creek (B.C.), 52
Rocky Mountains (Rockies), 5, 8,
15, 22, 23, 29-30, 148, 150, 152,
155, 156, 159
Rose Harbour (B.C.), 97
Royal City, 50
Royal Engineers, 7
Royal Geographical Society, 142,
163
Royal School of Mines, 45, 149,
154, 155
Royal Society, 142, 161
Royal Society of Canada, 142, 159,
162
Russians, 127, 129, 135-37

Sabiston, John (ship's captain), 94-
95, 98, 102
Salmon. *See* Fishing, salmon
Salmon House (B.C.), 68-71
San Francisco, California, 47, 76,
93, 94, 104, 157
Saskatchewan, District of, 13, 152,
153
Scotland, 153, 154
See also Edinburgh, Scotland
Seals and the seal fishery, 125, 127,
129-33, 134, 135, 136, 141, 161
Sea otters, 129
Selwyn, Dr. A.R.C. (director of
GSC), 5, 15, 48, 50, 58-59, 60-
61, 142, 148, 154, 155, 157, 159,
163, 164
Selwyn Inlet (B.C.), 99
Service, Robert (poet), 163
Shuswaps, 162
Shuswap Lake (B.C.), 84, 88, 157
Sicamous Narrows (B.C.), 86
Similkameen River (B.C.), 81, 89,
158
Sioux, 23, 27, 157-58
Skana, 63
Skidegate, B.C., 92, 95, 101
Skincuttle Inlet (B.C.), 99
Smallpox, 65, 147, 151, 152, 160
Snakes, 73, 88
Soda Creek, B.C., 50, 51
Steamers, 128-29
*See also Alaskan; Albatross;
Dakota; Danube; Ft. Lewis;
Grappler; Royal City; Skana*
Stewart River "troubles," 111, 119,
120
Stewart River (Yukon), 111
Stikines. *See* Tlingits
Stikine River (B.C.), 107, 111, 144
Strait of Georgia, 95, 150
St. George Island (Alaska), 129,
130

St. Lawrence River, 41, 42, 148
St. Paul Island (Alaska), 124, 129, 130
Sumallo River (B.C.), 81
Surveys, 5, 48
  See also British North American
  Boundary Survey Commission
Sweetgrass Hills. *See* Three Buttes

Tadoussac, Quebec, 41, 42, 152
Tanu, B.C., 99-101, 103
Tar Islands (B.C.), 101
Tatlayoko Lake (B.C.), 52, 54
Telegraph Creek, B.C., 110
Telegraph Trail, 72
Tenas men, 70-71
Tennyson, Alfred, Lord, 151, 152
Thompson, David (explorer), 4, 147, 148
Thompson Plateau (B.C.), 80
Thompson River (B.C.), 60, 80, 148, 157
Three Buttes (District of Alberta), 22-24
Tilley, Samuel (politician), 155
Tingle, Mr. (Alaska Commercial Company), 130, 131
Tlingits ("Stick"), 113-15
Tolmie, Dr. William Fraser (B.C. politician), 61, 149, 157, 160
Tommy (Dawson's crewman), 51, 53-56
Triangle Island, 125
Tsimshians, 113-17
Tuberculosis, 147, 159, 164
  See also Pott's Disease

Turtle Mountain, 9-10

Unalaska Island (Alaska), 128, 129, 138
University of Toronto, 142

Vancouver, B.C., 127, 138, 161, 164
Vancouver Island, 53, 150, 153, 155, 157, 160
Victoria, B.C., 47, 50, 61, 63, 76, 91, 93-94, 98, 105, 107, 155, 157

Waddington Party massacre, 54
Wanderer, 94, 97, 98, 105
Walkem, George (B.C. politician), 61, 156
Walrus, 134, 135
Washington, D.C., 139, 159, 162
Waterton Lakes (District of Alberta), 29, 30, 156
Westwold, B.C., 82
White Horse Rapids, (Yukon), 123
White Mud River (District of Alberta), 21
Williams (Dawson's crewman), 98, 102
Wrangell, Alaska, 111, 113

Yale, B.C., 50, 76
Yukon River, 111-12, 120-21, 122-23, 144
Yukon Territory, 108, 111, 112-23, 161, 164